Car electrical & electronic systems

Also from Veloce:

SpeedPro Series
4-Cylinder Engine Short Block High-Performance Manual – New Updated & Revised Edition (Hammill)
Aerodynamics of Your Road Car, Modifying the (Edgar and Barnard)
Alfa Romeo DOHC High-performance Manual (Kartalamakis)
Alfa Romeo V6 Engine High-performance Manual (Kartalamakis)
BMC 998cc A-series Engine, How to Power Tune (Hammill)
1275cc A-series High-performance Manual (Hammill)
Camshafts – How to Choose & Time Them For Maximum Power (Hammill)
Competition Car Datalogging Manual, The (Templeman)
Custom Air Suspension – How to install air suspension in your road car – on a budget! (Edgar)
Cylinder Heads, How to Build, Modify & Power Tune – Updated & Revised Edition (Burgess & Gollan)
Distributor-type Ignition Systems, How to Build & Power Tune – New 3rd Edition (Hammill)
Fast Road Car, How to Plan and Build – Revised & Updated Colour New Edition (Stapleton)
Ford SOHC 'Pinto' & Sierra Cosworth DOHC Engines, How to Power Tune – Updated & Enlarged Edition (Hammill)
Ford V8, How to Power Tune Small Block Engines (Hammill)
Harley-Davidson Evolution Engines, How to Build & Power Tune (Hammill)
Holley Carburetors, How to Build & Power Tune – Revised & Updated Edition (Hammill)
Honda Civic Type R High-Performance Manual, The (Cowland & Clifford)
Jaguar XK Engines, How to Power Tune – Revised & Updated Colour Edition (Hammill)
Land Rover Discovery, Defender & Range Rover – How to Modify Coil Sprung Models for High-performance & Off-Road Action (Hosier)
MG Midget & Austin-Healey Sprite, How to Power Tune – Enlarged & updated 4th Edition (Stapleton)
MGB 4-cylinder Engine, How to Power Tune (Burgess)
MGB V8 Power, How to Give Your – Third Colour Edition (Williams)
MGB, MGC & MGB V8, How to Improve – New 2nd Edition (Williams)
Mini Engines, How to Power Tune On a Small Budget – Colour Edition (Hammill)
Motorcycle-engined Racing Cars, How to Build (Pashley)
Motorsport, Getting Started in (Collins)
Nissan GT-R High-performance Manual, The (Gorodji)
Nitrous Oxide High-performance Manual, The (Langfield)
Race & Trackday Driving Techniques (Hornsey)
Retro or classic car for high-performance, How to modify your (Stapleton)
Rover V8 Engines, How to Power Tune (Hammill)
Secrets of Speed – Today's techniques for four-stroke engine blueprinting & tuning (Swager)
Sports car & Kitcar Suspension & Brakes, How to Build & Modify – Revised 3rd Edition (Hammill)
SU Carburettor High-performance Manual (Hammill)
Successful Low-Cost Rally Car, How to Build a (Young)
Suzuki 4x4, How to Modify For Serious Off-road Action (Richardson)
Tiger Avon Sports car, How to Build Your Own – Updated & Revised 2nd Edition (Dudley)
Triumph TR2, 3 & TR4, How to Improve (Williams)
Triumph TR5, 250 & TR6, How to Improve (Williams)
Triumph TR7 & TR8, How to Improve (Williams)
V8 Engine, How to Build a Short Block For High-performance (Hammill)
Volkswagen Beetle Suspension, Brakes & Chassis, How to Modify For High-performance (Hale)
Volkswagen Bus Suspension, Brakes & Chassis for High-performance, How to Modify – Updated & Enlarged New Edition (Hale)
Weber DCOE, & Dellorto DHLA Carburetors, How to Build & Power Tune – 3rd Edition (Hammill)

Workshop Pro Series
Setting up a home car workshop (Edgar)
Car electrical and electronic systems (Edgar)
Modifying the Electronics of Modern Classic Cars – the complete guide for your 1990s to 2000s car (Edgar)

RAC handbooks
Caring for your car – How to maintain & service your car (Fry)
Caring for your car's bodywork and interior (Nixon)
Caring for your bicycle – How to maintain & repair your bicycle (Henshaw)
Caring for your scooter – How to maintain & service your 49cc to 125cc twist & go scooter (Fry)
Efficient Driver's Handbook, The (Moss)
Electric Cars – The Future is Now! (Linde)
First aid for your car – Your expert guide to common problems & how to fix them (Collins)
How your car works (Linde)
How your motorcycle works – Your guide to the components & systems of modern motorcycles (Henshaw)
Motorcycles – A first-time-buyer's guide (Henshaw)
Motorhomes – A first-time-buyer's guide (Fry)
Pass the MoT test! – How to check & prepare your car for the annual MoT test (Paxton)
Selling your car – How to make your car look great and how to sell it fast (Knight)
Simple fixes for your car – How to do small jobs for yourself and save money (Collins)

Enthusiast's Restoration Manual Series
Beginner's Guide to Classic Motorcycle Restoration, The (Burns)
Citroën 2CV Restore (Porter)
Classic Large Frame Vespa Scooters, How to Restore (Paxton)
Classic Car Bodywork, How to Restore (Thaddeus)
Classic British Car Electrical Systems (Astley)
Classic Car Electrics (Thaddeus)
Classic Cars, How to Paint (Thaddeus)
Ducati Bevel Twins 1971 to 1986 (Falloon)
How to Restore & Improve Classic Car Suspension, Steering & Wheels (Parish – translator)
How to Restore Classic Off-road Motorcycles (Burns)
How to restore Honda CX500 & CX650 – YOUR step-by-step colour illustrated guide to complete restoration (Burns)
How to restore Honda Fours – YOUR step-by-step colour illustrated guide to complete restoration (Burns)
Jaguar E-type (Crespin)
Reliant Regal, How to Restore (Payne)
Triumph TR2, 3, 3A, 4 & 4A, How to Restore (Williams)
Triumph TR5/250 & 6, How to Restore (Williams)
Triumph TR7/8, How to Restore (Williams)
Triumph Trident T150/T160 & BSA Rocket III, How to Restore (Rooke)
Ultimate Mini Restoration Manual, The (Ayre & Webber)
Volkswagen Beetle, How to Restore (Tyler)
Yamaha FS1-E, How to Restore (Watts)

Veloce's other imprints:

www.veloce.co.uk

First published in November 2018 by Veloce Publishing Limited, Veloce House, Parkway Farm Business Park, Middle Farm Way, Poundbury, Dorchester DT1 3AR, England. Tel +44 (0)1305 260068 / Fax 01305 250479 / e-mail info@veloce.co.uk / web www.veloce.co.uk or www.velocebooks.com.
ISBN: 978-1-787112-08-7; UPC: 6-36847-01208-3
© 2018 Julian Edgar and Veloce Publishing. All rights reserved. With the exception of quoting brief passages for the purpose of review, no part of this publication may be recorded, reproduced or transmitted by any means, including photocopying, without the written permission of Veloce Publishing Ltd. Throughout this book logos, model names and designations, etc, have been used for the purposes of identification, illustration and decoration. Such names are the property of the trademark holder as this is not an official publication. Readers with ideas for automotive books, or books on other transport or related hobby subjects, are invited to write to the editorial director of Veloce Publishing at the above address. British Library Cataloguing in Publication Data – A catalogue record for this book is available from the British Library.
Cover photograph courtesy Andrey Popov / iStock.
Typesetting, design and page make-up all by Veloce Publishing Ltd on Apple Mac. Printed in India by Parksons Graphics.

Car electrical & electronic systems

JULIAN EDGAR

WORKSHOP PRO CAR ELECTRICAL AND ELECTRONIC SYSTEMS

INTRODUCTION 6

1 CAR ELECTRICITY 8

2 SWITCHES AND RELAYS 16

3 MULTIMETERS 26

4 FAULT-FINDING BASIC CAR ELECTRICAL SYSTEMS 36

5 ANALOG AND DIGITAL SIGNALS 50

CONTENTS

6 USING ELECTRONIC COMPONENTS 60

7 OSCILLOSCOPES 72

8 ENGINE MANAGEMENT 84

9 OTHER CAR ELECTRONIC SYSTEMS 122

10 FAULT-FINDING ADVANCED CAR SYSTEMS 134

11 ELECTRONIC BUILDING BLOCKS 146

APPENDIX 160

INDEX 166

Introduction

The greatest change in automotive technology over the last 40 years has been the integration of electrical and electronic technologies into every aspect of a car's workings. You could argue that this particular process began long before 40 years ago (think: electric starting, electric ignition and electric lighting), but it's in more recent times that the number of electronic control systems has just exploded. As described later in this book, some cars now have over 30 different electronic systems – from those that control the engine, to those that operate the seats, to those that operate the door locks.

So, you might be thinking, how can a single book cover all of that? The answer is that it cannot, but I've taken the approach best described in the proverb '*give someone a fish and you feed them for a day; teach them to fish and you feed them for a lifetime.*' What I have tried to do is to give you enough background that not only can you fix problems in your car as they appear, but you will be in a good position to understand electrical and electronic systems that you've not seen before.

It's for this reason that I start with the simplest of all circuits – a light and a battery. From there we talk about different types of circuits (eg parallel and series), and introduce the ideas of voltage, current and resistance. Without these, understanding electrical (and later, electronic) systems is like trying to write without knowing the alphabet.

From there, we look at how to use switches and relays. I find simple circuits using switches and relays great fun, and I've used them extensively over the years in car modification. (Once, I developed a modification that switched off the traction control, but left the Electronic Stability Control active, all achieved using just relays to switch signals.) The immobiliser covered in Chapter 2 is one of the simplest but most effective ways I've seen of preventing your from car being stolen – especially an older car without a high level of security.

In Chapter 3 we look at how to use a multimeter. Being able to confidently measure voltage, current and resistance is vital if you are to work on car systems. From there we dive straight into fault-finding within the types of electrical systems that all cars have: lights, starting systems, charging systems and ignition. If you're confident with these older electrical systems, and can already use a multimeter, relays and switches, I'd suggest you read Chapter 1 and then move straight to Chapter 5 – Analog and Digital Signals. Covering these signals might seem a big jump from what we've been talking about so far – but it isn't. Here we also meet technologies like communication buses (eg CAN bus) and explore ideas like frequency and duty cycle.

In Chapter 6 I look at how electronic components – parts like resistors, capacitors, diodes and transistors – work in circuits. Together with a good understanding of the concepts covered in the first chapter, this content will allow you to understand how many larger electronic systems operate. Even if you do not build any of the circuits covered in this chapter, seeing how they work will let you better diagnose faults and work on a range of car systems. (So, what's a pull-up resistor, then – and why is it needed?)

Earlier, we covered the use of a multimeter, and in Chapter 7 we introduce its big brother – the oscilloscope. One of the brilliant aspects of changing technology is that oscilloscopes (scopes) have dropped greatly in price, making a scope affordable for anyone working on a car. A scope provides a window into how many electronic systems work – it literally draws you pictures of what you would otherwise not be able to see or measure. In this chapter, I use nearly 20 different scope patterns to show you what those pictures look like.

We then head into the biggest chapter in this book – engine management. Electronic Fuel Injection (EFI), and then later engine management (ie control of both fuel and spark), had the earliest penetration of all car electronic systems into the marketplace. As a result, it is these car electronic systems where people spend most of their time. I've chosen to cover engine management using Bosch systems as examples, from L-Jetronic right through to ME-Motronic. I've done this for two reasons: first, because Bosch is the only engine management manufacturer that makes available really high quality technical material on their systems, and second, because almost every engine management system and component in the world follows the Bosch approach to a large degree. Understand Bosch systems, and you'll understand most others.

In Chapter 9 we meet some of the other car electronic systems, like ABS and climate control. In this chapter, I also cover a specific approach to car electronics that will allow you to quickly get your head around systems with which you're initially not familiar. Then, in the next chapter, we look at fault-finding more complex car systems. Thankfully, the advent of On Board Diagnostics (OBD) has made this far easier than it once was, but for those people who have pre-OBD cars, I also talk about some approaches that will work in those situations.

Finally, I cover some of the electronic building blocks that you can buy off the shelf and easily integrate into your car. This is a great approach to car modification – or even just to improve car convenience. The electronic modules are cheap, pre-built and, in addition to power and ground connections, typically need only a few wires connected. That makes it much easier to implement a simple but effective project that doesn't involve disappearing into a nightmare of wiring!

I've been working on car electrical and electronic systems all my life. They're fascinating and fun – and I hope that this book helps you realise both of those aspects.

FURTHER READING

So what do you read when you've finished this book? The best resources available on car electrical and electronic systems are those from Robert Bosch GmbH. The company was founded in 1886 and in almost every area of car electrical and electronic systems, has lead the world ever since. (And when hasn't it? The company lagged well behind when hybrid cars were introduced – but that's another story.) Bosch publishes a range of material, including the best overall guide in this area – the Automotive Handbook. The company also publishes books on specific areas of electronics – diesel engine management, safety systems, and so on. Bosch publications are not for beginners (and they can be quite dense), but if you have read and understood the content of this book, you should be fine.

The other resources you should pursue are those published by the company that made your car. In-house technical materials (typically designed to introduce dealer mechanics to the new systems) are often excellent. The main difficulty is getting hold of these materials, but often helpful people publish the material on the web. If you have an older car, the manufacturer's official workshop manual is also worth buying.

I mention this later but it's worth reiterating here. When researching this book, I was surprised at the amount of very useful material published by makers of scan tools. For example, Snap-on has available for free download excellent material on accessing fault codes (including those that do not need a tool) and general fault-finding.

Finally, if you become really fascinated with car electronic systems and where they are going, the US-based Society of Automotive Engineers publishes a large number of technical papers each year – and many of these are on electronic systems. Be warned though – first, the papers are expensive to buy, and second, they are written for fellow engineers and can be very complex.

Julian Edgar

WORKSHOP PRO CAR ELECTRICAL AND ELECTRONIC SYSTEMS

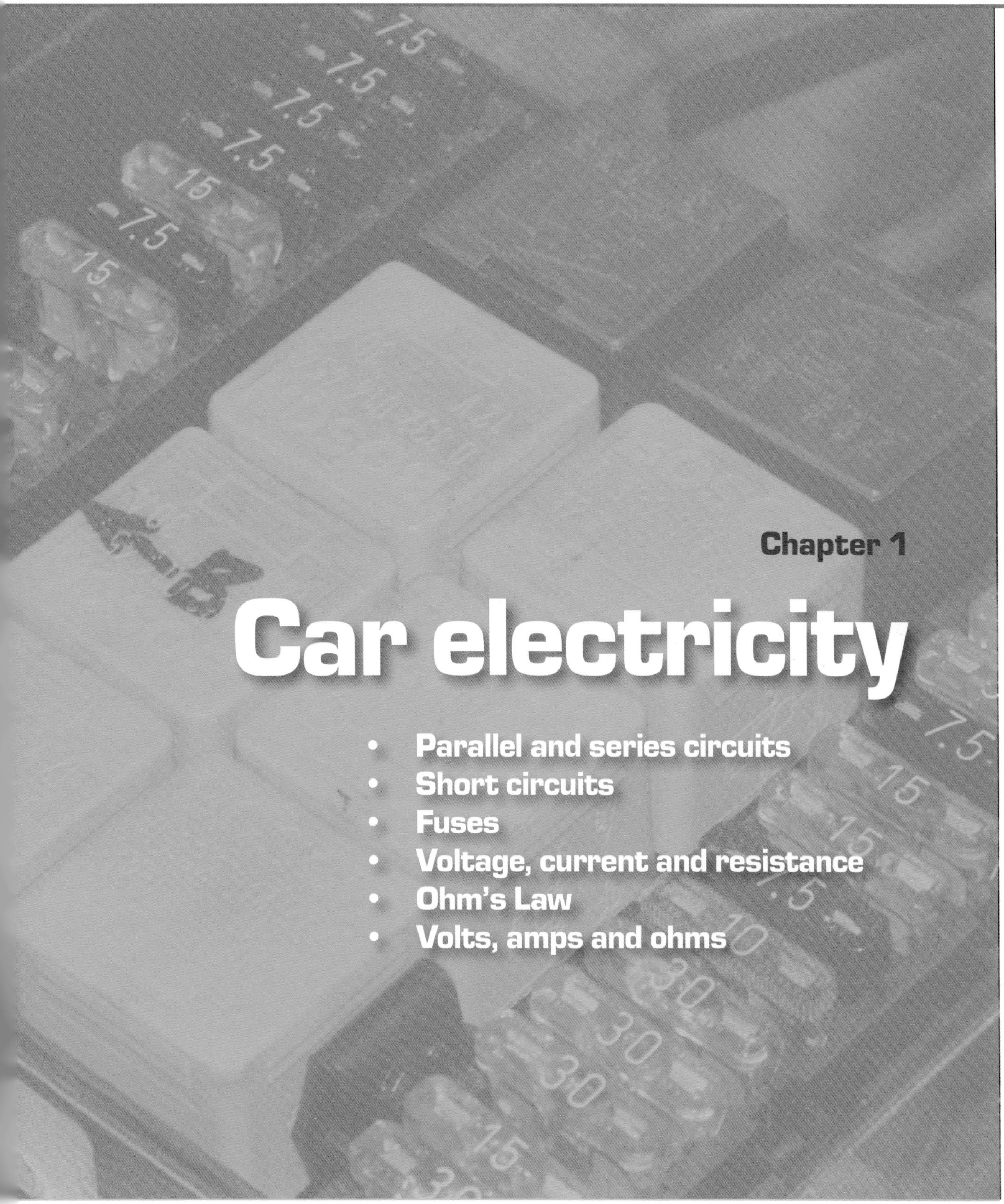

1. CAR ELECTRICITY

Chapter 1

Car electricity

- Parallel and series circuits
- Short circuits
- Fuses
- Voltage, current and resistance
- Ohm's Law
- Volts, amps and ohms

WORKSHOP PRO CAR ELECTRICAL AND ELECTRONIC SYSTEMS

When working with car electrical and electronic systems, you need to first understand some fundamental ideas – eg what is a circuit, what are some of the different types of circuits, and what are voltage and current and resistance. So let's explore those ideas and then immediately apply them to cars.

CIRCUITS

All electrical systems that we will deal with in cars are made up of circuits. Rather than attempt to define an electrical circuit, it's easiest to describe the operation of one. After I've done that, I'll look at parallel and series circuits, and then describe how you can use the idea of circuits in car electrics.

Picture a 12 volt (V) car battery out of the car and sitting on a bench. It has two terminals, one marked positive (+) and the other negative (-). If we connect a 12 volt light bulb to the battery, one terminal of the light bulb to the positive terminal and one to the negative terminal, the light will glow. This is shown in Figure 1-1.

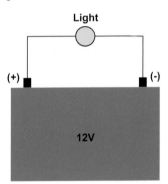

Figure 1-1: A complete circuit is needed before electricity will flow and, in this case, the light work.

The bulb glows because *the circuit is complete*. Electricity can flow from the positive terminal, through the filament of the light bulb and then back to the negative terminal. If we break the circuit anywhere, the light goes off. That break could be a gap in the wiring, a broken filament in the bulb, or an open switch. A break in the circuit is shown in Figure 1-2, and the light is off. So, for a circuit to work, the electricity needs to be able to find its way from the battery positive terminal all the way around and back to the negative terminal.

In cars, the negative terminal of the battery is connected to the car's bodywork. This connection is called a 'ground connection.' Because the car is made of metal, it acts as a very large wire. Therefore, instead of one side of the light bulb being connected directly back to the negative terminal of the battery, it can instead be 'grounded' – that is, connected somewhere to the car's metal body. ('Ground' is sometimes called 'earth.') The metal body acts as the wire that connects the ground back to the negative terminal of the battery. Figure 1-3 shows this approach. We've all heard of a 'bad ground' being responsible for headlights flickering or an engine running poorly – you can now see that if this connection is intermittent, the circuit will at times be being broken.

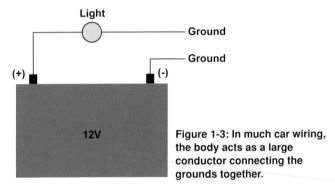

Figure 1-3: In much car wiring, the body acts as a large conductor connecting the grounds together.

PARALLEL AND SERIES CIRCUITS

Circuits can be organised in two basic ways. These ways are called *series* and *parallel*.

Imagine again the battery connected to the light bulb, as before. This time, though, we're going to add a second light bulb. If we break the existing circuit and insert the second light bulb, the electricity will now take the following path: electricity flows from the positive terminal of the battery through the filament of the first light bulb, then through the filament of the second light bulb, and then back to the negative terminal of the battery. This is shown in Figure 1-4.

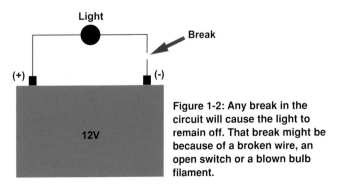

Figure 1-2: Any break in the circuit will cause the light to remain off. That break might be because of a broken wire, an open switch or a blown bulb filament.

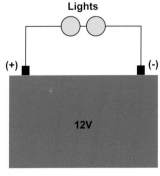

Figure 1-4: These lights are connected in series – electricity flows through one light, and then the other.

1. CAR ELECTRICITY

A circuit designed like this is a *series circuit* – the electricity flows through one device and then through the next. In a series circuit of two light bulbs, if one light fails, both lights will go off. In the same way, a switch placed anywhere in the circuit will turn off both lights. Figure 1-5 shows a single switch operating both lights.

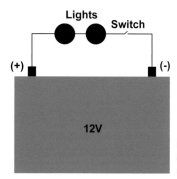

Figure 1-5: In a series circuit, any break (such as this open switch) will turn off both lights.

We can also wire the two light bulbs to the battery in a different way. Instead of wiring them one after the other, let's *independently* connect them to the battery. To do this, we connect the first light bulb to the battery (one terminal to positive and one terminal to negative), and then we do just the same for the second light bulb. This is shown in Figure 1-6. This type of circuit is called a *parallel circuit*. Figure 1-7 shows the same circuit drawn in a slightly different way. You can see why this is called a parallel circuit – the two lights are arranged parallel to one another.

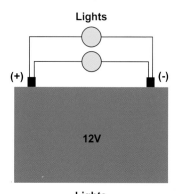

Figure 1-6: These lights each have their own connections to the battery. They are said to be wired in parallel.

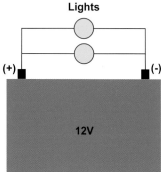

Figure 1-7: This wiring approach is commonly used in parallel circuits, with single connections to the battery feeding multiple circuits.

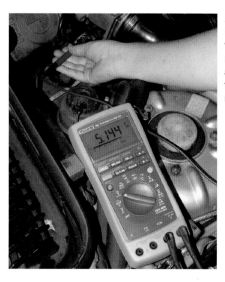

A multimeter is inserted in parallel with the circuit when measuring voltage, and in series with the circuit when measuring current.

In a parallel circuit of two light bulbs, if one bulb fails, the other is not affected. In the same way, a switch in one bulb's circuit will not influence the other bulb. Most circuits in a car are parallel circuits. In a parallel circuit, the electricity has separate paths for each load, while in a series circuit, the electricity has to successively flow through all loads. (This means that in a series circuit, each powered device sees a voltage that is lower than battery voltage – more on voltage in a moment.)

Series and parallel circuits are used in every branch of electrical and electronics work, from the simplest to the most complex. The behaviour of voltage, current and resistance are all affected by whether the circuit is a series or parallel design.

SHORT CIRCUITS AND FUSES

When the electricity can take a 'short-cut' straight from the positive to the negative terminals, and so does not have to pass through any electrical component that normally takes power, there is said to be a *short circuit*. Figure 1-8 shows a short circuit. In this case, not only would the light bulb go out (the electricity would take the short cut rather than going through the bulb), but so much electricity would

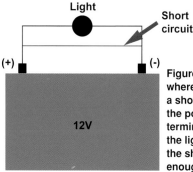

Figure 1-8: A short circuit is where electricity can take a short-cut straight from the positive to the negative terminals. In this situation, the light would go off and the short circuit may get hot enough to start a fire.

11

Fuses protect circuits from excessive current flows of the sort that happen if a short circuit occurs. Here a 7.5 amp blade fuse is being removed for inspection. The square boxes are relays, which are covered in the next chapter.

flow in the circuit that a fire would probably start. To avoid the risk of fire caused by a short circuit, a fuse is used to protect circuits. A fuse is just a thin piece of wire that melts if too much electricity flows through it.

Over the years, different types of fuses have been used in cars. Cars of 30 and 40 years ago often used glass fuses, where the thin wire was contained inside a glass tube. Other cars used ceramic fuses, where the thin wire ran along a short length of ceramic. Most current cars use blade fuses, where the fuse wire is inside a plastic assembly. Blade fuses come in two different sizes. In all cases, a fuse works through the excess current heating a thin wire until it melts.

(Some cars have circuit breakers instead of fuses. In this approach, a circuit breaker is like a switch that turns off when the current flow is excessive. Circuit breakers can be manually reset if they have tripped.)

Let's use fuses to show how series and parallel circuits are often used together in a car. Figure 1-9 shows a headlight circuit, complete with fuse and closed switch. See how all the electricity that powers the headlights must pass through the fuse – the fuse is in *series* with the headlights.

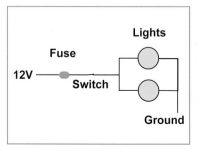

Figure 1-9: Here are both series and parallel circuits. The fuse is in series with the lights – if the fuse blows, both lights go out. But the headlights are wired in parallel – if one headlight bulb blows, the other headlight stays illuminated.

But look at how the headlights are wired in *parallel* with one another. We can test the series/parallel approach we took previously to see if this is the case. Remember, in a series circuit, any break in the circuit turns off the load. Let's imagine the fuse blows. Do the lights turn off? Yes; so the fuse is in series with the headlights. Now, let's imagine one headlight bulb blows. Do both headlights go off? No, so the headlights are wired in parallel with each other.

PARALLEL AND SERIES CIRCUITS IN USE

In cars, the ideas of parallel and series circuits are applied all the time. So that you can see how important these ideas are, let's take a look at some more examples.

Adding a second horn

Let's say that you are unhappy with how loud your car's existing horn is. You decide that you want to add a second one just like the first, and wonder how to wire the new one into place. There are two wires going to the existing horn. You could either:

1. Cut one of the wires going to the original horn and insert the new horn in this wire (giving a series circuit).

Or

2. Bare a section of each of the wires going to the original horn and then connect the new horn to each of those wires (giving a parallel circuit).

Approach 1 (the series circuit) won't work, because the horns will each get only 6 volts, not the 12 volts they need. Approach 2 (the parallel circuit) must be used.

Hybrid and electric car batteries

High voltage battery packs used in electric and hybrid cars use lots of low-voltage cells wired in series. For example, one Toyota Prius pack uses 240 nickel metal hydride cells (each 1.2 volts) wired in series to give an output voltage of 288 volts. (You can think of this as each cell pushing the electricity along by another 1.2 volts. So after the first cell the voltage is 1.2 volts, after the second cell it's 2.4 volts, and so on.) If the cells were wired in parallel rather than series, the voltage output would be only 1.2 volts. So depending on the wiring configuration, the output voltage is either 1.2 volts or 288 volts! That's a huge difference.

Using a multimeter

I will cover the use of a multimeter in more detail in Chapter 3, but at this stage it's important to know that when using a multimeter, some measurements are taken with the meter in *parallel* with the circuit, and some are taken with the meter in *series* with the circuit.

For example, if you are measuring the output of an engine's airflow meter, you 'ground' the negative probe of

1. CAR ELECTRICITY

the multimeter and connect the other probe to the output terminal of the airflow meter. The meter is being used in *parallel* with the airflow meter circuit. (How can you confirm that? Well, if you disconnect the multimeter, the circuit from the airflow meter to the Electronic Control Unit [ECU] isn't broken.)

On the other hand, if you want to measure how much current is flowing in a headlight circuit, you need to break the circuit and insert the meter in it. Often this is easiest done by temporarily pulling a fuse and touching the multimeter probes to each side of the fuse holder. Now the meter is being used in *series* with the headlight circuit. (Again, how can you confirm that? Well, if you disconnect the multimeter, the circuit is broken and the headlights go off.)

VOLTAGE

A circuit, that we looked at earlier, consisted of only a battery and a light. The battery was marked as being '12V' or 12 volts. But what does this mean? Like many electrical terms, it's easiest to understand if an analogy is used – the voltage of electricity is a bit like *pressure* – for example, the pressure of fuel in a fuel-line. A fuel pump in a car pressurises fuel, pushing it through the fuel line to the injectors. A battery produces an 'electrical pressure,' causing an electric current to flow through a circuit. Electrical pressure is measured in volts.

The higher the voltage, the greater the distance that an electrical spark will jump. The ignition system produces a voltage of more than 20,000 volts, and this high voltage allows the spark to jump across the plug's electrodes.

If you measure the voltage of a car battery with the engine stopped, the voltage might be 12.5 volts. However, with the engine running and the alternator charging the battery, the voltage might increase to 13.8 volts or even more. In this case, measuring the battery voltage, firstly with the engine stopped, and then secondly with the engine running, shows us that the alternator is working.

In the same way, voltages in car systems are often used as *indicators of something happening*. For example, an oxygen sensor in an exhaust has an output voltage that varies with mixture strength. With a narrow-band sensor, the voltage might be 0.2 volts when the mixture is lean, and 0.8 volts when the mixture is rich. In this case, the engine management ECU measures voltages coming from the sensor to work out mixture strength.

Car system voltages of around 12 volts are not dangerous – you cannot get an electric shock from a normal car battery. As described above, ignition systems, however, use much higher voltages and so can give you a shock. Hybrid and electric cars use high voltage batteries and are *very dangerous*. It's for this reason that on these cars, the cables that carry these high voltages are sheathed in orange tube.

CURRENT

Current is the *amount of electricity flowing past a point*. Using the fuel line example, it's like measuring how many litres (or gallons) per second are passing along the pipe. Current is measured in amps. Wires that need to take a lot of current (like the one to the starter motor) are thick.

Fuses are a good example through which to understand current. As I said above, a fuse uses a very thin wire inside it – so thin that if the fuse is required to flow a current greater than its rating, the wire melts and breaks the circuit. For example, a fuse for a car radio might be rated at 5 amps. The radio draws less than 5 amps, and so the fuse does not blow. However, if the wire that supplies the radio rubs against a body bracket and the insulation wears through, a short circuit could develop and the current flow would be much higher than 5 amps, causing the fuse to blow. And what if you were to connect a large car sound amplifier to the same radio wiring? In that case, the amplifier will draw a lot more than 5 amps, and so again the fuse will blow. In that situation, you need to connect the amplifier directly to the car battery with a new, high-current fuse.

In the following table are some examples of current flows in an operating car. (These are just indicative and vary from car to car.)

Item	Current (amps)
LED interior light	0.25
Low beam headlights	7
Radiator fan	15
Powerful car sound amplifier	35
Maximum alternator output	120
Starter motor	200

Note that any device that contains coils and magnets takes a big current gulp when first switched on. These devices include motors, solenoids and fuel injectors. Solenoids and injectors also develop voltage spikes when they're switched off, something we will come back to later.

RESISTANCE

Resistance is a measurement of *how hard or easy it is for a current to flow through a substance*. Something with a really high resistance is called an insulator – it lets almost no current through it. On the other hand, anything which allows current to flow very easily is called a conductor. The normal copper wires within a car loom are good conductors, while the plastic covering them is a good insulator – stopping the current from going where it's not intended to. As the resistance goes up, the flow of electricity is reduced (more on this relationship in a moment). There are lots of gradations between good conductors and good insulators, and the exact value of the resistance posed to the flow of electricity is measured in ohms.

Many engine management sensors vary in their resistance. For example, the coolant temperature sensor is a resistor that varies in electrical resistance with temperature. Here is a table that shows measured resistance of a particular coolant temperature sensor at different temperatures:

Temperature (degrees C)	Resistance (ohms)
0	6000
20	2200
40	1200
60	600
80	350
100	190
120	110

Resistors can also be thought of as being a bit like a restrictor on a pipe. I mentioned earlier that you can think of voltage as being rather like fuel pressure, and current flow as being like fuel flow along the pipe. If you put a restrictor in a pipe, that will be a pressure drop across the restrictor. In the same way, if a resistor is placed in a circuit, there will be a voltage drop across the resistor. And, again like the flow of fuel along a pipe, the greater the current flow, the greater the voltage drop across the resistor.

Driving lights are rated in watts, so how can you work out how many amps the supply cable has to handle? Watts divided by volts equals amps, so if you know the lights' wattage, finding the current is an easy calculation.

VOLT-OHM-AMPS RELATIONSHIPS

There is a strict mathematical relationship between voltage, current and resistance. (It's called Ohm's Law.) For example, if you know the voltage drop across a known value of resistance, you can work out how much current must be passing through the resistor.

The equation is:

$$amps = volts \div ohms$$

This can be rearranged as:

$$volts = amps \times ohms$$

and

$$ohms = volts \div amps$$

There is also another relationship with which you should be familiar.

It's this:

$$volts \times amps = watts$$

'Watts' (W) in electrical terms has the same meaning as 'watts' (or kilowatts) when applied to car engines – it's the rate at which work is being done.

This equation can also be expressed as:

$$amps = watts \div volts$$

and

$$volts = watts \div amps$$

Unless it's easy for you, don't bother memorising all these – just remember that there are strict relationships between volts, ohms, amps and watts. Change one, and the others change by an exact amount.

USING THESE RELATIONSHIPS

Let's look at an example of these relationships, starting with voltage, resistance and current.

Figure 1-10 shows a circuit of the sort with which we're now familiar. However, rather than the lights we've been powering, in this case we've inserted into the circuit a resistor, a device that, as the name suggests, resists the

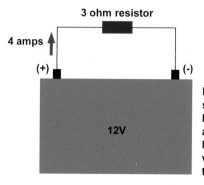

Figure 1-10: Ohms Law shows the relationship between voltage, current and resistance. If you have any two of the values, you work out the third.

1. CAR ELECTRICITY

flow of electricity. With Ohm's Law, we need only two of the three variables and then we can work out the other one. So in this circuit, 12 volts divided by the 3 ohms resistance equals 4 amps of current flow. Or, 4 amps of current flow multiplied by 3 ohms of resistance means that there is 12 volts doing the 'pushing.' Or, 12 volts divided by 4 amps of current flow equals 3 ohms of resistance.

Let's look at an example using watts. You decide you want some driving lights on the front of your car. They're rated at 50 watts each, and because you are using two, you know you'll need to supply enough power to run 100 watts of extra lighting. You go off to find some suitable wire but you find that automotive wire isn't rated in watts; instead it's rated in amps. So how many amps will need to flow in this new wire? Watts ÷ volts = amps. We know the wattage is 100 watts. We know that car systems run on (about) 12 volts. So what is 100 watts divided by 12 volts? It's 8.3 amps. Use 10 amps cable and you'll be fine.

Exactly the same idea applies to more powerful systems. A hybrid petrol/electric car might have a 30 kilowatt electric motor and a battery pack that provides 288 volts. So how much current does the wiring need to handle? Watts divided by volts equals amps, so that works out to 30,000 watts (30 kilowatts) divided by 288 volts, which is over 104 amps! No wonder the cables are so thick ...

Now what about a trickier example? Let's say that someone has told you that your trailer brake lights are very dim. You check and they're working – but the person was right, they *are* very dull. You get out the multimeter, unplug the trailer and probe the socket on your car. When someone puts their foot on the brake pedal, there's a measured 12 volts at the socket – so that's OK. *Or is it?*

Here's a clue: remember I said previously that a resistance in a circuit is a bit like a restriction in a fuel pipe? If the current flow (ie amps) is very low, the voltage drop across the resistance is also very low. A multimeter takes only a tiny amount of current from the circuit it is measuring, so if there is any resistance in the circuit, measuring voltage won't show it. You need to have lots of current flowing to see what is really happening.

Now, instead of unplugging the trailer and measuring at the socket, this time leave the trailer plugged in and make the same measurement. This time we're measuring the voltage *under load* – and what do you know, the voltage is now only 8 volts! There must be a dodgy connection that is causing a resistance. Without the correct current (amps) flow, the voltage drop across the resistance was not visible.

Let's look at the idea of voltage drops in more detail. Previously I said that volts = amps x ohms; therefore, the higher the current (amps), the greater the voltage drop for a given resistance. So? Well, take the example of where you have a really large amount of current flowing. A starter motor can easily take 200 amps when the engine is cranking, and the starter motor is normally fed straight from the battery via a short length of thick cable. Because the cable is short, the voltage drop is fairly low – let's call it 0.5 volts. That is, on cranking there might be 12.5 volts at the battery and 12.0 volts at the starter motor.

But now let's move the battery to the other end of the car, perhaps to create more room in the engine bay. Rather than 1 metre (about 3ft) of cable between the battery and the starter motor, there might now be 5 metres (about 15ft) of cable. The battery cable has a certain fixed resistance per length, so if we increase its length by five times, the resistance is five times greater. We had a 0.5 volt drop with the shorter cable, so with five times the length, we now have (5 x 0.5 =) a 2.5 volt drop. With the revised battery location, and using the same size cable, we now have the starter motor seeing only 10 volts – and that's too low. This is why when a battery is moved to the other end of the car, really thick connecting cable is needed.

(And how much resistance does the battery cable have in the above example? The voltage drop was 0.5 volts per metre of cable at a current flow of 200 amps. Remember that ohms = volts ÷ amps, so here we have 12.5 volts divided by 200 amps, which equals 0.063 ohms per metre of cable.)

UNITS

We've talked about volts and amps and ohms, but in many automotive uses, different prefixes to these units are used. An example of this is to describe a resistance as 10 kilohms, meaning 10,000 ohms. The table below shows some of the prefixes with which you should be familiar. It also shows the abbreviations for volts, amps and ohms – and from here on in this book we will use these abbreviations.

Basic unit	Units for small amounts	Units for large amounts
Volts (V)	millivolt (mV) $1mV = 0.001V$	kilovolt (kV) $1kV = 1000V$
Amps (A)	milliamps (mA) $1mA = 0.001A$	kiloamps (kA) $1kA = 1000$ amps (This unit is rarely used in car applications)
Ohms (Ω)	milliohms (mΩ) $1m\Omega = 0.001\Omega$ (This unit is rarely used in car applications, but may be used to describe the resistance of heavy cables.)	kilohms (kΩ) $1k\Omega = 1000\Omega$ (This unit is frequently written as 'kilo-ohm.')

WORKSHOP PRO CAR ELECTRICAL AND ELECTRONIC SYSTEMS

Chapter 2

Switches and relays

- **Switch specifications**
- **Switch circuits**
- **Switch types**
- **Identifying switch terminals**
- **Relays**
- **Relay specifications**
- **Using relays**
- **Solid-state relays**

WORKSHOP PRO CAR ELECTRICAL AND ELECTRONIC SYSTEMS

In the previous chapter, we saw that a circuit needs to be complete for a device to operate. We also saw that the circuit can be broken by a switch. In this chapter, I want to look at switches, and then their electro-mechanical equivalent, relays. You can achieve a lot by the smart use of switches and relays, but especially when they develop problems, these devices can also be a source of frustration and difficulty.

A 12V rocker switch that illuminates when turned on.

SWITCH SPECIFICATIONS

All switches make or break circuits. That is, they have contacts that when the switch is operated, are connected together or moved apart.

The simplest switch is an on/off switch for a single circuit, as shown in Figure 2-1. In this circuit, closing the switch turns on the light; opening the switch turns off the light. Note the symbol that is used for the switch. The arrow is called the 'pole' and the direction it can switch is called the 'throw.' In this case, there is only one pole that can go in only one direction to activate the output, so the switch is called a Single Pole Single Throw (SPST) switch. SPST switches have only two contacts, and it doesn't matter which way around these are connected in a circuit.

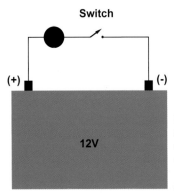

Figure 2-1: The simplest switch is an on/off switch, as shown in this circuit. This type of switch is called a Single Pole Single Throw (SPST) switch.

A coolant temperature switch of the type often used to activate radiator fans.

Let's move on to the next switch type. Imagine we want to control two different devices with the one, two-position switch. In the 'up' position we want to operate one device, and in the 'down' position we want to operate another device. An example of this is a single switch that allows us to select a loud horn for country use and a quieter horn for city use. Figure 2-2 shows this type of switch in the circuit for the two horns. Note how the switch still has a single pole (one arrow) but now that arrow can be switched in two different directions (ie it can activate two different outputs) – the switch has two throws. This type of switch is called a Single Pole Double Throw (SPDT) switch, and has three connecting terminals.

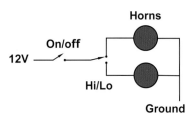

Figure 2-2: A circuit that allows you to select between two different horns (eg loud and quiet). The switch on the right that selects the different horns is called a Single Pole Double Throw (SPDT) switch. The other switch is the horn switch.

The next useful switch type to consider is a Double Pole Single Throw type. As a single throw switch, this design has only one position where an output is activated. However, because it is a double pole design, it can switch two totally different circuits at the same time. For example, you might want to turn on a dashboard warning light whenever a device is operating. One pole of the switch can operate the device (eg a pump), while the other pole turns on the dashboard light. The two circuits can be completely separate, which in many situations makes the wiring easier. Figure 2-3 shows this circuit. Note how this switch has four terminals. Note also that the red bar on the switch indicates that both contacts move together – you can think of this as an insulating connecting bar.

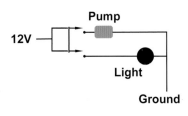

Figure 2-3: The switch in this circuit is called a Double Pole Single Throw (DPST) switch. It allows you to simultaneously switch two completely different circuits. Here a dashboard light is illuminated when you switch on a pump.

18

2. SWITCHES AND RELAYS

A Double Pole Double Throw (DPDT) toggle switch. The two middle terminals are each connected to the adjacent top or bottom terminals, depending on the position of the toggle.

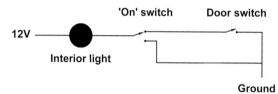

Figure 2-5: Using two switches in series to control an interior light. The 'on' switch will illuminate the light all the time, while when this switch is in the other position, opening the door will turn on the light.

Now that we've got used to idea of analysing switch types on the basis of the number of poles and throws that a switch has, we can picture a:
- Double Pole Double Throw (DPDT – six terminals)
- Single Pole Triple Throw switch (often abbreviated to 1P3T – four terminals)
- Single Pole Four Throw (1P4T – five terminals)

Switches with three or four (or more) throws are often of a rotary design, where you turn a knob to different click positions. This is because a toggle or slide switch with this many positions would get rather hard to operate.

In addition to the switches described above, there are switches available with many other combinations of poles and throws.

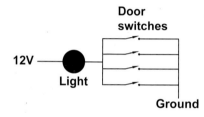

Figure 2-4: This circuit shows how switches can be wired in parallel. Four door switches are able to operate the interior light – open any (or all) doors, and the light will illuminate.

SWITCH CIRCUITS

Let's take a look at some switches in operation in a car.

The interior light in an older car is directly operated by door switches. Usually, each door switch connects to ground when the car door is opened. The switches are therefore SPST switches – but with four or five of them wired in parallel. If any (or all) of the switches are closed, the light will be on. Figure 2-4 shows this circuit with four door switches.

Let's stay with interior lights. Say you want to add an interior light in the rear load area of an older hatchback car that doesn't have one. You want it to come on when you open the hatch but you also want to be able to manually select an 'always on' position. Figure 2-5 shows the circuit, that uses two switches. A door switch closes when the hatch is opened. However, this switch comes into operation only when selected by the SPDT switch. When the SPDT switch is in the other position, the light is on all the time.

Especially in older cars that do not use as many electronic control modules, some switches can be quite complex. For example, Figure 2-6 shows the wiper/washer switch of a late 1990s Honda. It uses two switches to control the wipers, the one on the left for wiper speed and the one on the right to operate the 'mist' (single wipe) function. The wiper speed switch is a three pole three throw (3P3T) design. (Note the red insulated bar that shows the connection between the throws.) The positions give intermittent, slow and fast speeds. The mist function uses a double pole single throw (DPST) switch.

The approach taken by Honda to show multi-pole, multi-throw switches is shown in Figure 2-6, but not all manufacturers take this approach in their circuit diagrams. Figure 2-7 shows the hazard lights switch of a Toyota. Note how a different diagram approach is taken, with bars showing which terminals are linked with the switch in the 'off' position, and which terminals are linked with the switch in the 'on' position.

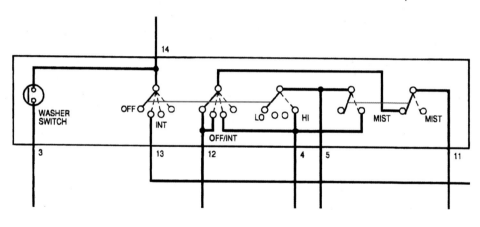

Figure 2-6: This windscreen wiper switch uses two switches to control the wipers, the one on the left for wiper speed and the one on the right to operate the 'mist' (single wipe) function. The wiper speed switch is a three pole three throw (3P3T) design. Note the red insulated bar that shows the connection between the throws. The four positions give intermittent, slow and fast speeds. The mist function uses a double pole single throw (DPST) switch. At far left is a simple switch for the washer pump. (Courtesy Honda)

19

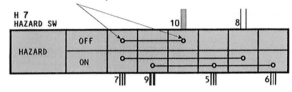

Figure 2-7: Here a switch is drawn using a different approach. Bars show which terminals are linked with the switch in the 'off' position, and which terminals are linked with the switch in the 'on' position. (Courtesy Toyota)

SWITCH TYPES

In addition to the number of poles and throws, switches come in different types. We've mentioned switches that are toggle (which use a lever) and rotary (which use a rotating knob). But in addition, there are also rocker switches, pushbutton switches and many other types.

Some switches are momentary – these switches resume their normal position when released. Most momentary switches are 'normally open' (they close the circuit only when operated) but some momentary switches are 'normally closed' – they open the circuit when operated. Momentary pushbutton switches are good examples of those that can be bought as either normally open or normally closed. A door switch is typically a normally closed momentary switch – it's the closing of the door that pushes (and holds) the switch open.

Switches are also rated in terms of their current and voltage. In nearly all car applications, the voltage ratings of a switch will be fine – 12V is a low voltage for most switches to handle. On the other hand, it's easy to exceed the *current* rating of a switch. For example, in Chapter 1 we looked at the current requirement for a pair of 50W spotlights, and calculated that we needed wiring that could cope with 8.3A. To directly switch these lights, you'd need a switch rated for at least 10A – so, a heavy-duty switch.

Electric motors take a large gulp of current when first switched on – for example, a radiator fan might have a short-term gulp of 25A, settling down to 15A when the fan is running. These are large current loads – too much for a small switch. (So how do you operate such a fan? With a relay – more on relays in a moment.)

Some other switches are available that operate automatically when certain physical conditions are reached. For example, there are temperature switches that automatically close at 40°C (104°F) and open at 35°C (95°F). I use these sorts of switches (in normally open form) to control fans in car sound amplifiers. An oil pressure switch opens when engine oil pressure exceeds a designated level. These switches are used to turn on a dashboard warning when the oil pressure is low. Some older cars with fuel-injection used an inertia switch that turned off the fuel pump if the car was subjected to a collision or was upside-down.

When selecting a switch for a new application, a few steps will help you select the right switch:
- Sketch the circuit to determine how many poles and throws you will need the switch to have.
- Decide on the switch type (eg pushbutton, rocker, rotary, momentary, etc).
- Especially if it is a pushbutton, does it need to be normally open or normally closed?
- What current rating is needed?

Note that in many cases, a switch being sold for general use in electronics is more appropriate than a specific automotive switch.

IDENTIFYING SWITCH TERMINALS

The more poles and throws that a switch has, the harder it is to identify what switch terminals do what. On the other hand, a SPST switch is easy – there are just two terminals and these can be connected either way around in a circuit. In a SPDT switch, there are three connections, normally arranged in a line. The centre of the three connections is the pole that can be switched to the two other connections. A DPDT switch is similar to a SPDT design, except the there are two rows of three terminals next to one another. However, rotary and unusual switch designs may not be as straightforward as the switches so far described. In those cases, use the 'continuity' function on a multimeter (see the next chapter) or a power supply and test light to work out which terminal is which.

Some switches incorporate lights. Sometimes, these lights turn on when the switch is activated, while in other designs, the light is designed to be illuminated at night as part of the dashboard lighting. If the light is part of the dashboard lighting, there will be two terminals for it – these are often smaller, or a different colour, to the main switch terminals. Operating this type of light is as simple as connecting power to it. If, on the other hand, the light is designed to turn on when the switch is activated, there will probably be only one terminal for the light. Again, this terminal will be marked in some way. This terminal is normally connected to ground, power is connected to one of the other terminals and the load to the final terminal. Note that all switches with lights in them must be rated for 12V DC.

RELAYS

Relays are switches that are operated by an electromagnet. The electromagnet (or coil) has only two terminals. When these terminals are connected to power, the electromagnet is activated and pulls across the switch contacts. You will able to hear a click as this occurs. Most

2. SWITCHES AND RELAYS

A Single Pole Double Throw (SPDT – sometimes also called 'changeover') automotive relay. Note the diagram on top of the relay, which shows the connections, and the pre-wired plug, which makes installation easy.

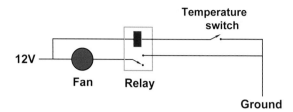

Figure 2-8: A low current temperature switch is being used to trigger a relay that in turn operates a radiator fan. The green box shows the relay, with the blue rectangle indicating its coil.

relays use coil contacts that can be connected either way around, but some are polarised as they use a diode to protect a transistor that might be feeding power to the relay. (More on this in Chapter 6.)

Because they are simply switches, relays can be categorised on exactly the same basis as switch contacts. A SPST relay has just one contact that is closed when the relay is activated. A DPST relay has two contacts that are closed when the relay is activated. If you understand switches, relays are easy – just remember that a relay will always have an extra two terminals for the coil.

On automotive SPST relays, the pins are given standardised numbers. The coil connections are 85 and 86, while the two connections for the internal switch are 30 and 87. However, most general-purpose relays don't have any numbers on the pins – instead the functions of the pins are usually shown on a little diagram on the body of the relay.

The most common use for a SPST relay is to use a small electrical current to control a large electrical current. For example, remember how I said previously that a radiator fan takes a lot of current, and a small switch wouldn't be able to handle it? As a real-world example, it is cheap and effective to use a simple temperature switch to control radiator fans – but such a switch won't handle the current. But what if you used a relay in-between? The temperature switch then needs to handle only the small current required by the relay's coil, while the switching of the fan itself is handled by the relay's high-current contacts. This circuit is shown in Figure 2-8.

Another example of where a relay is used to allow a small current to control a large current is when you are using an electronic module to control a device. Most electronic modules won't be able to directly drive high-current loads (eg a siren, pump or fan) and instead drive a relay that does the hard work. (We will look at some of these modules in Chapter 11.)

SPST relays used as standard fitment in cars usually perform the function of allowing a small current to control a large current. For example, there might be relays for headlights, the horn, the fuel pump, the air compressor clutch, the radiator fan – and so on.

A SPDT relay (sometimes called a 'changeover' relay) allows you to control two devices, for example switching one off as the other is switched on. An example of where I needed to use a relay in this way is in a fuel system that needed to be switched between two different fuel pressures. To raise the fuel pressure, a solenoid valve had to be turned off and, at the same time, a fuel pump needed to be switched on. Both devices drew a fair amount of current so a heavy-duty automotive relay was used. This circuit is shown in Figure 2-9.

Automotive SPDT relays use the following codes for their pins: the coil connections are again 85 and 86, the normally open output is 87, the normally closed output is 87a, and the input is 30.

A DPDT relay allows you to switch two different circuits simultaneously. With this type of relay, you can:
* Turn on two completely independent circuits
* Turn one off and one on
* Turn off two completely independent circuits

These relays are less common in automotive aftermarket use and so don't have coded numbers for the pins.

So what use is a DPDT relay, then? Again, I'll use an example from a car modification I did. What was needed was the on-demand disconnection of two oxygen sensor input signals to the ECU. The two signal wires from the

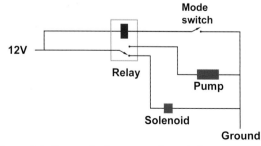

Figure 2-9: To vary fuel pressure, I needed to switch off a solenoid and at the same time, activate an additional fuel pump. A SPDT relay has been used to achieve that.

21

WORKSHOP PRO CAR ELECTRICAL AND ELECTRONIC SYSTEMS

A 12V Double Pole Double Throw (DPDT) relay. Note how six of the terminals are organised as we saw earlier with the DPDT switch – they work in just the same way as the switch. The two extra terminals are for the coil. It's fun looking through the transparent case and seeing the relay contacts move when power in applied to the coil.

oxygen sensors to the ECU needed to be kept completely separate, so they couldn't be joined together and a SPST relay used. Instead a DPDT relay was used. (It didn't actually have to be a double throw design, but DPDT relays are more common than SPDT designs.)

All relays with double throw designs have some contacts that are normally closed, and some contacts that are normally open. We're familiar with these ideas from switches, but in a relay it's important to know that the normally closed contacts *are in that state when the relay coil is not powered.* That is, when the relay is energised:
- The normally closed contacts go open circuit
- The normally open contacts go closed circuit

While relays can be used to switch large currents with small currents, another very important function of a relay is to *invert the function of a circuit*. For example, when something is switched off, another item is automatically switched on. It's not an automotive example, but it's the very first time I ever considered using a relay; as a 12-year-old, I wanted to set up a burglar alarm (good for catching brothers and sisters!). I wanted a circuit where, when a wire was broken (eg by a switch opening as a window was raised), it triggered a siren. So when one circuit was turned off, it turned on another circuit. A relay easily achieves this inversion of logic – see Figure 2-10.

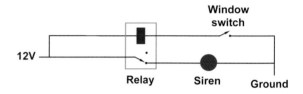

Figure 2-10: Here are relay has been used to invert logic. In this simple household burglar alarm, lifting a window opens the switch, causing the SPDT relay to switch off. The siren is therefore activated.

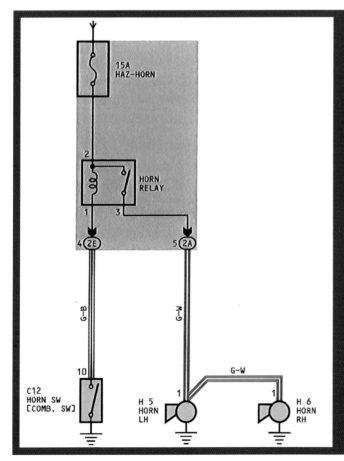

CIRCUIT DIAGRAM SYMBOLS

There are no standardised symbols for switches and relays on car circuit diagrams. For example, the symbol that is used for a relay may have the coil shown as a curly piece of wire, as a rectangular empty block, or as a square with a diagonal line in it. The relay contacts may be shown in the same way the manufacturer draws switches, or they may look quite different.

The trick when reading circuit diagrams is to not look for particular symbols, but to always remember that a switch will have contacts that open and close, and a relay will have both a coil and contacts. For both relays and switches, work out: what are the poles and what are the throws? Which contacts are normally open and which are normally closed? Which part of the relay is the actuating coil? Identify these first and you will understand the circuit, no matter how it is drawn.

In this Toyota circuit diagram, the horn relay is drawn with the coil indicated by a curly wire, and the relay switch contacts shown in the same way as a normal switch. However, not all manufacturers use this approach. (Courtesy Toyota)

And an automotive example? I recently set up an air suspension system that used a compressor and a solenoid valve block. Whenever a particular valve was open in the valve block, the compressor had to be switched off – otherwise, the compressor just pumped air out through this open valve into the atmosphere. It sounds complex, but by adding a normally closed relay with the coil wired in parallel with the valve, and the contacts wired in series with the compressor, it meant that whenever the valve opened, the compressor was disabled.

Small relays are often used when signals (low currents) are switched – more on this later.

One interesting use of a relay is as a 'latching relay.' A latching relay is a normal DPST (or DPDT) relay that is wired so that it stays on ('latched') after it has been triggered. This triggering can be by a momentary, normally open button. Often, the circuit is unlatched by pressing another, normally closed button, or by switching off power to the relay. Using two relays (one wired to be a latching relay), it's easy to make an immobiliser that requires a very special trick before you can start the car. Figure 2-11 shows such a circuit.

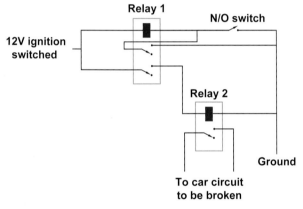

Figure 2-11: This immobiliser circuit uses two relays and a single normally open (N/O) pushbutton switch. The circuit is shown in the 'car immobilised' position – note how Relay 2, which is in series with the fuel pump or ignition, is open-circuit, so the car won't run. When we want the car to go, we press the normally open switch. This applies power to Relay 1's coil, so pulling the contacts up. In turn, this allows power to continue being fed to the coil, even when we release the switch; that is, this relay is now latched. The second set of contacts on Relay 1 feed power to the coil of Relay 2, so pulling its contacts up and feeding power to the fuel pump or ignition. When we switch off the ignition, Relay 1 automatically unlatches. Now here's the really good trick. Let's make the switch a reed switch – a switch that closes only when there is a magnet nearby. To make the car go, we then have to swipe a magnet past the hidden reed switch, that can be located behind a plastic part of the dashboard. I used to hide the magnet inside a normal-looking car remote, so even if the thief steals your keys, they still cannot drive your car away – not unless they know the special 'swiping' trick!

Almost buried under the wiring loom are some of the engine bay relays and fuses in this Mercedes.

I always enjoy using relays. They're durable (very hard to blow up), straightforward to wire, and can achieve quite tricky outcomes in car electrical systems.

RELAY SPECIFICATIONS

In addition to its contact configuration (SPST, DPDT, etc) there are at least three other relay specifications that are important:

Coil voltage

As its name suggests, coil voltage refers to the voltage which the relay coil works on. All car relays are designed to work on 12V. A nominally 12V relay is fine on car voltages, even though these voltages can extend as high as 14.4V. However, in many cases, you will want to use relays not designed specifically for car use. For example, when switching low current signals, general purpose relays are cheaper and come in many more configurations than automotive relays. In those cases, ensure that you select relays that have 12V coils. For example, you shouldn't use a 5V coil relay on a 12V system.

Coil current

This is the amount of current the relay coil will draw when energised. This can be expressed directly in milli-amps (thousandths of an amp), or indirectly as a coil resistance.

A very sensitive relay might have a coil resistance of 360Ω. Assuming a running-car voltage of 13.8V, 13.8V divided by 360Ω gives a coil current of 0.038A, or 38mA. In other words, the switch that you're using to operate the relay has to handle just 38mA. This is a very low value of required current. A typical automotive relay is more likely to have a coil resistance of 80Ω, giving a coil current flow of 170mA. (13.8V ÷ 80Ω = 0.17A) If you have the relay to hand, it's easy to work out how much current the coil will draw by measuring the resistance with a multimeter and then doing the calculation.

Maximum contact current

This spec refers to the max current that a relay's contacts can handle. To avoid arcing, you should use a factor of safety so that the maximum current of your switched circuit (even when it first switches on with a current gulp) is less than the relay's spec. Automotive relays are available with current ratings such as 25A, 30A and even 60A. Be careful when checking maximum current specs that the listing is for the DC (direct current) at or above the voltage you'll be using – ie, in cars, 13.8V. A relay rated at 10A at 240V AC, for example, is not the same as one rated at 10A at 12V DC.

Standard car relays (those with the square enclosure and spade terminals) vary substantially in internal quality. I remember once when I'd fitted a pair of powerful driving lights to my car, and I was out testing them at night. Much to my consternation, when faced with oncoming traffic, the dazzling lights would not turn off! The relay had stuck in the 'on' position. When I pulled the relay apart, I found that the contacts had been manufactured so they were not properly parallel with one another, and so when the relay was switched on, the high current passed through just a tiny contact surface area. At that point, the contacts had melted, fusing them together. It's worth buying quality car relays.

Other specs that you might find listed for relays include life (ie how many millions of operations the relay will do before failure) and perhaps response time – the later seldom important in automotive uses.

USING RELAYS

Using a relay is made a lot simpler if you follow these steps:
- Draw a circuit diagram. The first step is to draw a simple circuit diagram showing where the wires go. Which wires go to the relay coil, which to the normally open and normally closed contacts of the relay?
- Decide what type of relay is needed. If just one connection needs to be switched on and off, you'll use a SPST design. If two connections need to be switched, a DPST or (more commonly) a DPDT design will be the one to use. A changeover (where one device is switched off and the other switched on) can use a SPDT or a DPDT design.
- Work out the functions of each pin. If it's a standard automotive relay, read the numbers. If it's a general-purpose relay, look for the diagram on the relay body. If neither of these apply, by careful use of a short-circuit protected power supply and a multimeter, you can work out the functions of each pin. (Unless you use too high a test voltage, you can't damage the relay.)
- Wire the relay coil first. If you wire the relay's coil first, you'll be able to check that the relay is working by listening to its click.

A solid-state relay like this can repeatedly switch high currents without wear. These relays use a large transistor to do the switching, rather than mechanical contacts.

SOLID-STATE RELAYS

If you are frequently switching high current loads, you may wish to use a solid-state relay. Solid-state relays replace the electromagnet and moving contacts of a mechanical relay and instead use an electronic switch (a big transistor). They are available only as SPST designs. As you'd expect with a SPST relay, there are four connections – two on the low current side, and two on the power side. However, unlike a conventional mechanical relay, all the connections on a solid-state relay are polarised – they must be connected to negative or positive as indicated.

When buying a solid-state relay, ensure that it is a DC:DC relay, not the more common DC:AC designs that are used to control mains AC power. The relay should also have a maximum continuous current rating that is sufficient – for example, 40A continuous. If you look at the specs, you'll

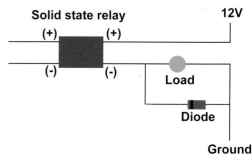

Figure 2-12: If you are using a solid-state relay to switch inductive loads like electric motors, you will need to protect the relay against the voltage spikes using a diode.

2. SWITCHES AND RELAYS

see that a relay rated at this continuous current is good for two or three times this value for a short time, allowing it to easily cope with the big switch-on current gulp of the load. If in doubt, uprate the relay's current capacity.

If you are switching an inductive load (one with coils and magnets), you will need to protect the solid-state relay against the voltage spikes that come from these devices when they are switched off. You can achieve this with a diode wired across the device, as shown in Figure 2-12. (More on diodes in Chapter 6.)

RELAY BOARDS

Standard car relays are frequently used by those adding a circuit to a car, and of course are used if replacing a standard plug-in car relay that has failed. But what if you are designing a new circuit? In that case, one of the cheap relay modules now available (eg on eBay) may be better suited to your purpose.

These relay modules integrate high current relays, screw-type connecting terminal strips for the high current and low current connections, and often have a LED relay on/off indication incorporated on the board. Additionally, many of these relay modules use transistor control and so take only minuscule currents, meaning that the activation switches can also be tiny, low current designs.

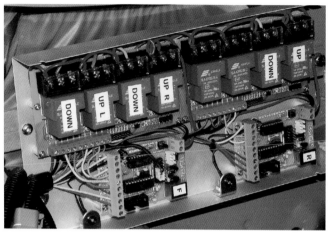

Lots of relays! Here, a pair of quad-relay boards have been used as part of the control system for a custom air suspension system. There are relays for the front-left, front-right and rear suspension. (Two relays are unused.) The relays are each SPDT designs (note the three terminals for each relay on the upper terminal strips) and are switched by transistors triggered by the two control boards (lower). The relays are used to operate air solenoids that control airflow in and out of the air springs. Using pre-built relay boards can save time and effort when you need to use multiple relays in a custom design, or you need to trigger the relays with only very small currents.

WORKSHOP PRO CAR ELECTRICAL AND ELECTRONIC SYSTEMS

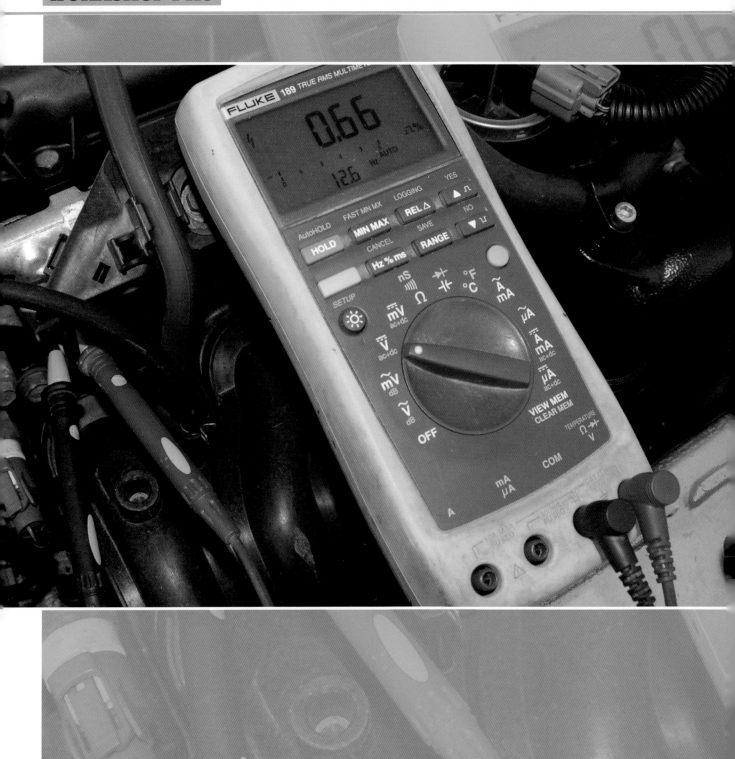

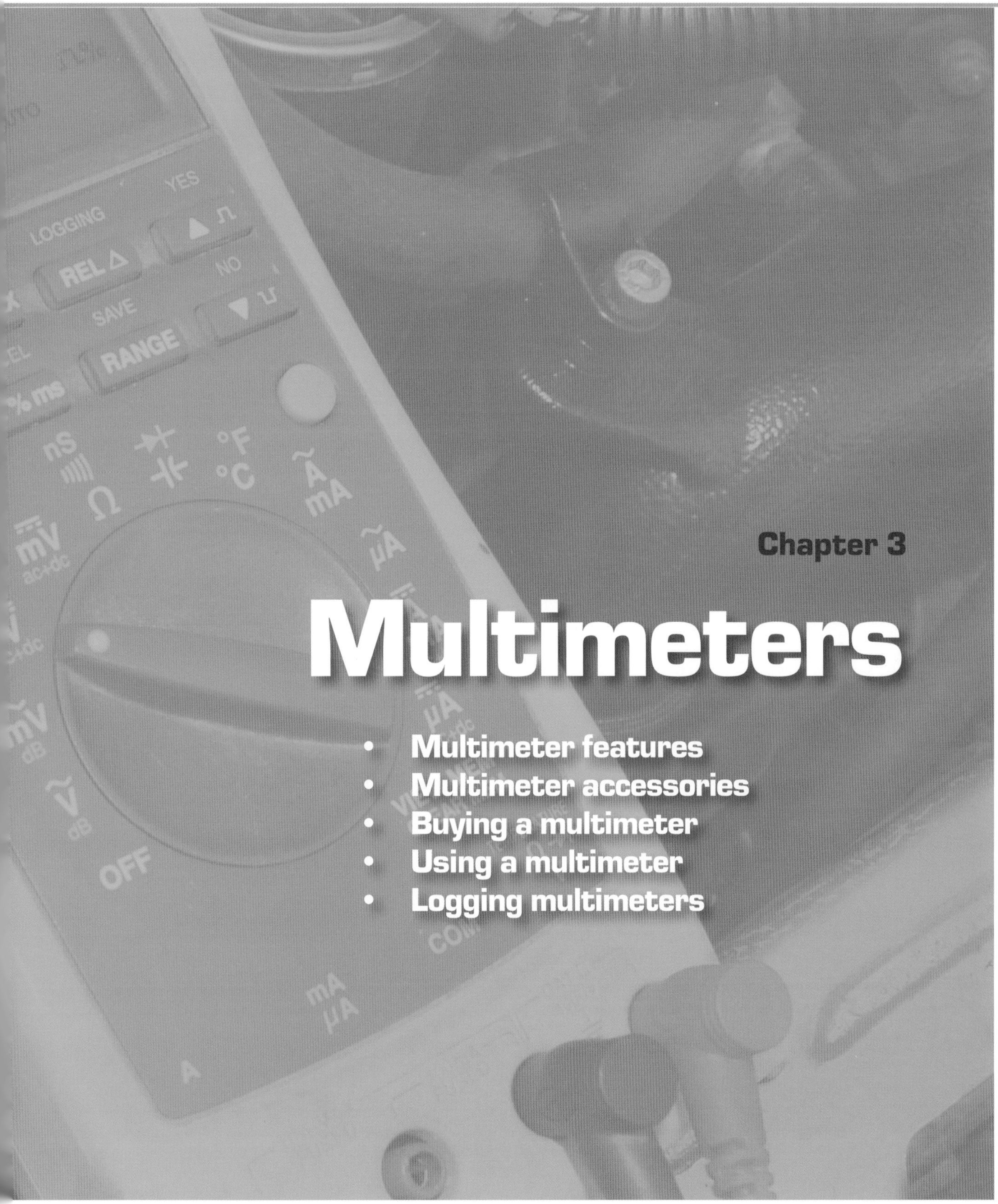

Chapter 3
Multimeters

- **Multimeter features**
- **Multimeter accessories**
- **Buying a multimeter**
- **Using a multimeter**
- **Logging multimeters**

WORKSHOP PRO: CAR ELECTRICAL AND ELECTRONIC SYSTEMS

To do any effective work on the electrical and electronic systems in your car you'll need to use a multimeter. In this chapter, I'll start by looking at the features of a multimeter and then move on to its use. In doing this, we will be building on the content of the first two chapters – so, if you're not yet familiar with parallel and series circuits, and switches and relays, best start there first!

Left: A low-cost multimeter. Even this meter can measure frequency and duty cycle, in addition to voltage, current and resistance. Right: A high-quality multimeter – I've used one like this for many years. The backlit display is useful in dark footwells.

MULTIMETER FEATURES

A multimeter is a test tool which can measure a variety of different electrical factors – at minimum: volts, ohms and amps. In addition to these measurements, it is also useful if the meter displays:
- Continuity
- Frequency
- Duty cycle
- Temperature (via a plug-in probe)

A meter with max/min/average functions is even more useful. So what are these extra features, and why are they important?

A continuity function sounds a beeper when the meter's probes are connected together. If the probes are connected to either end of a wire, the sounding of the beeper shows that there is no break in the wire; that is, it has continuity.

Frequency refers to how often something is changing – in a car, that usually means being switched on and off. For example, many flow control valves are rapidly pulsed – an idle speed control valve, or a turbo boost control valve. Being able to measure frequency is important if the valves are to be controlled by a new system (for example, programmable engine management) or if you are fault-finding.

Duty cycle refers to the proportion of time a device is switched on for. For example, a fuel injector at idle speed might have a duty cycle of only 2 per cent – the injector is open and flowing fuel for only 2 per cent of the time. Especially in a modified car where power has been increased, being able to read the maximum duty cycle is important – at 100 per cent duty cycle, the injectors are working at full capacity. Duty cycle measurements are also used when fault-finding, for example of an idle control valve.

Temperature measurement is via a K-type thermocouple that can be plugged into the meter. Thermocouples are typically available as bead designs (these are small and lightweight, which allows their temperature to change very rapidly, but they are fragile) or probe types (much more rugged but their temperature changes more slowly). Often, it's worth having one of each sensor type – bead designs are easy to slip under hose clamps, for example. You can use probes to measure the temperature of the coolant, engine and gearbox oil, and intake air.

Many meters have a 'peak hold' or similar function. This displays the maximum reading that occurred during the measuring period. This is especially useful if you are testing on the road and cannot safely watch the meter. For example, for best performance, an engine should breathe cold air. If you are making a new intake, you can use a temperature probe and a multimeter to find which areas in the engine bay get hot, and which stay cool. Moving the probe around and using the peak hold function on the meter will soon give you this information.

More expensive meters not only show the maximum reading that was gained, but can also show the minimum and average readings as well. If the meter can sample very rapidly, these additional functions are very useful. For example, I used the output of a height sensor to measure the effectiveness of a front spoiler on a car. If the spoiler was being effective, you'd expect that ride height would decrease (or at least not get higher!) as the car went faster. However, on a bumpy road, the output of the sensor is constantly changing, so it's hard to make sense of the reading. What is wanted in this situation is a reading of average height. This allows the ride height to be compared at the same speed with and without the spoiler fitted. To make this measurement, I used my high-speed averaging multimeter.

Multimeters must have what is called a 'high input impedance.' This means that when you apply the meter to the system that you are measuring, the meter won't

An auto-ranging multimeter like this requires only that you select the right parameter – the meter then sets the correct range based on the value being measured.

3. MULTIMETERS

Having minimum and maximum functions is useful in many measuring situations. This meter can also show the average reading of the signal.

draw more than a tiny amount of current. Meters that don't have a high input impedance (old analog meters and some cheap digital meters) will load down the system. For example, measuring the output of a narrow-band oxygen sensor will be impossible with a low impedance multimeter – and attempting to do so may well damage the sensor. When looking at meter specs, the meter should have an input impedance of at least 10 Megohms.

Different multimeters have a different number of digits on their LCDs. This can be readily seen by looking at the catalog picture, or at the meter in the flesh. But what you see may be not what you're actually getting – there's a trick involved in understanding what the meter can actually show you.

A typical low-cost multimeter has what is called a '1999 count.' That is, it has four digits with the last three digits able to display all numbers from 0-9, but the first digit able to be only '0' (sometimes blanked) or '1.' The highest number that can therefore be displayed is 1999 (or 1.999, 19.99, 199.9). Confusingly, this type of display is often also called a '3½ digit' display – the '½' indicating that the first digit is capable of showing only '1' or '0.' Next up the sophistication list are '3999 count' or '3¾ digit' designs. These have a maximum display number of 3999, 3.999, 39.99, or 399.9. Really top meters go as high as '50,000 count' or '4 ⅘ digits' and can display numbers like 50000, 5.0000, 50.000, 500.00, 5000.0. It's easy to get lost in all of this, but just remember: the higher the 'count' or 'digit' number, the more detail you can read.

Multimeters are available in auto-ranging or manual-range types. An auto-ranging meter has much fewer selection positions on its main knob – just amps, volts, ohms and temperature, for example. When the probes of the meter are connected to whatever is being measured, the meter will automatically select the right range to show the measurement.

Meters with manual selection must be set to the right range first. On a manual meter, the 'volts' settings might include 200mV, 2V, 20V, 200V and 500V. When measuring 12V battery voltage in a car, the correct setting would be '20V,' with anything up to 20 volts then able to be measured.

While an auto-ranging meter looks much simpler to use – just set the knob to 'volts' and the meter does the rest – the meter can be slower to read the measured value. This is because it first needs to work out what range to operate in. If the number dances around for a long time before settling on the right one it can be difficult to make quick measurements, and even more difficult if the factor being measured is changing at the same time as well! However, to speed up readings, some auto-ranging meters also allow you the option of fixing the range. Note that expensive multimeters will very quickly get the right reading, even if they are auto-ranging.

A backlight function is very useful when working with cars – it allows nighttime on-road testing and also makes things easier when working in darkened footwells.

Some meters have two displays, although they still have only one pair of input leads. The two displays are used to simultaneously show two characteristics of the one signal that's being measured. To do this, the two different signals have to be on the same input – you can't show temperature and voltage for example. But if you are measuring (say) a pulsed solenoid that controls turbo boost, you can simultaneously measure both its duty cycle and frequency – one display shows duty cycle and the other, frequency. But in most car applications this isn't all that advantageous – you usually only want the one parameter measured at a time.

For automotive use, look for a meter design which comes in a brightly-coloured rubber holster – it helps protects the meter from damage as well as making it easier to find – and one which is protected against the entrance of moisture. Good meters use 'O'-rings to seal the case and jacks.

Leads

Multimeters come with leads that are equipped with sharp probes. These are fine for general purpose measurements but for best car use, you should buy some additional accessory leads. I suggest that you buy extra leads equipped with the following:

- Alligator (crocodile) clips – they are useful when one side of the multimeter needs to be grounded, eg by connecting to a chassis bolt.
- Very sharp insulating-piecing probes – these allow you to tap into a wire without it being disconnected from the circuit. These probes are easily damaged, so take good care of them.

WORKSHOP PRO CAR ELECTRICAL AND ELECTRONIC SYSTEMS

A set of accessory probes and adaptors for a multimeter will save you a lot of time and effort.

- Spring hook probes that use miniature hooks that will lock around terminals, eg the terminals inside an injector socket, allowing the measurement of injector resistance.

Buy leads that are silicone insulated as they'll be more durable than leads with conventional insulation.

Current clamp

A current clamp is a multimeter accessory that allows you to measure much higher current flows than a normal multimeter can handle. A current clamp outputs a precise voltage per measured amp. For example, it might have an output of 1 millivolt per amp (mV/A). This makes measuring the clamp's output easy – if the multimeter shows a measurement of 5mV on its voltage scale when connected to the operating clamp, the current flowing in the wire is 5A. If the voltage displayed on the multimeter is 100mV, the current flowing in the wire is 100A.

When using a current clamp, its jaws are opened, the clamp passed over the wire, and the jaws closed. The wire is then centred in the opening and the measurement made. Note that it's the *individual conductor* that is measured – not a cable containing both earth and power leads, for example.

Current clamps are not particularly good at accurately measuring very small currents. This is so for two reasons. Firstly, if the output scale of the clamp is 1mV/A, a current flow of 0.5A is only 0.5mV – a figure that is very low for many multimeters to accurately measure. In addition, because of the influence of stray magnetic fields, current clamps need to be zeroed before they can be used. That is, a knob on the clamp first needs to be turned until the current reading is zero – obviously, when there isn't any current flowing through a wire inside the jaws! Getting the meter to read precisely zero can be fiddly. For these reasons, current clamps are usually used for current measurements of about 5A and upwards. (Most multimeters have a maximum current rating of 10A, so in practice the overlap between a current clamp and a multimeter works fine.)

If you want to be able to directly measure the current draw of high-current devices like starter motors, air suspension compressors, car sound amplifiers, electric seat motors and the like, you need a current clamp. The same also applies when measuring alternator output.

Pressure sensor

Pressure sensors are available that will plug into a multimeter. The sensor can be used to measure fuel pressure, intake manifold pressure (both positive and negative), oil pressure and so on. The benefit of using an electronic sensor over a mechanical pressure gauge is that, depending on the meter being used, you may be able to measure not only the 'live' value but also the maximum, minimum and average values. A fast-response pressure sensor can also be used with an oscilloscope – more on scopes in Chapter 7. Like current clamps, pressure sensors output a certain voltage per unit of pressure. The Fluke PV350 sensor I have can be switched to either metric or Imperial units, and outputs a voltage of 1mV DC per unit.

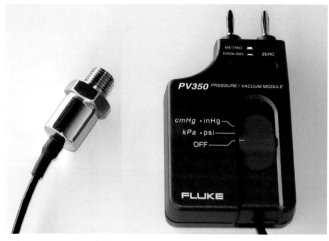

A pressure measuring attachment for a multimeter. Used with a good multimeter, with peak-hold and similar functions, this accessory is very useful.

3. MULTIMETERS

BUYING A MULTIMETER

If you are new to car electronics, buy a low-cost multimeter – for example one that covers just voltage, current and resistance measurements, and has a continuity function. However, if you want to be able to use functions of the other sorts mentioned above, then you'll be up for a much bigger outlay. In that case, I suggest that you buy a good quality meter. I use a Fluke multimeter for my main measurements and a second, cheaper multimeter when I want to monitor two circuits simultaneously. I also have a Fluke current clamp, pressure sensor and thermometer adaptor. (However, I don't buy Fluke probes and leads – they are much too expensive for what you get!)

USING A MULTIMETER TO MEASURE VOLTS

When measuring all but current flow (amps), the multimeter is applied in *parallel* with the circuit. For example, if you want to measure the voltage at a headlight, the negative probe of the meter is grounded (ie connected to body chassis) and the other side is connected to the power supply at the headlight. The meter is set to 'volts DC' to make this measurement. In most voltage measurements carried out on a car, the negative probe of the meter is grounded and the other probe connects to the signal of interest.

Measuring voltages is carried out when diagnosing almost every system in a car: measuring signal voltages being outputted from sensors, measuring that a device is being fed the correct power supply voltage to allow it to work, measuring battery voltage when the battery is being charged.

Let's look at another example of measuring voltages. You suspect the throttle position sensor (TPS) on an older car is faulty. As we will cover in more detail later, the throttle position sensor has three connections – ground, a regulated 5V, and the signal output. When the throttle is moved, the signal output should smoothly change from around 0.8V-4.5V. (The actual maximum and minimum might be different to these values – these are indicative.) You'd also expect that the signal output voltage would smoothly change with throttle movement.

The first step is to access the wiring. In an older car that's often easiest done at the plug on the TPS itself. You carefully pull back the rubber boot so that you can back-probe the plug – that is, get electrical access to the plug while it is still connected. Set the meter to DC volts and connect the negative probe to ground. You could clip it to the negative terminal of the battery (if that's handy) or to a plated (not painted) chassis bolt.

Turn on the ignition switch and then apply the positive probe of the multimeter to the terminals of the plug, one at a time. For example, the first terminal might have 5.02V on it. It doesn't change when the throttle is moved – so that's the regulated 5V feed to the sensor. The next terminal has 0V on it – that's ground. The final terminal has 0.85V on it, and when the throttle is moved, this voltage increases, reaching a maximum of 4.3V – this is the signal output.

However, even though the throttle is being moved smoothly, the voltage on this terminal doesn't rise smoothly. Instead it appears to suddenly drop to zero at about 25 per cent throttle, before going back to normal as the throttle is moved further. Issue? The TPS is indeed worn – get a replacement.

It's important to know that if back-probed in this way, and with the multimeter set to read volts, you cannot damage any car system. However, you *could* damage the system if your probe slipped and bridged, for instance, the 5V and ground connections in the plug, so you always need to be careful.

Measuring voltages is the most common use of a multimeter on a car. Note that are there are four settings for volts – AC and DC volts, and volts and millivolts. You will almost always want DC volts.

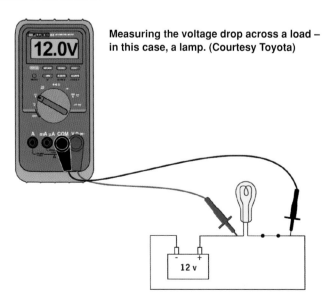

Measuring the voltage drop across a load – in this case, a lamp. (Courtesy Toyota)

In many measurements, the negative probe of the multimeter needs to be connected to ground. A large clip makes this easy.

Back-probing an airflow meter plug with a multimeter.

Measuring the signal reading of the airflow meter.

In addition to measuring voltages with respect to ground (that is, with one probe grounded), voltages can also be measured in a different way. If you wish to measure the *voltage drop* between two points, simply connect the two probes to those two points. I say that this approach is different because one of those points might not be ground.

For example, in Chapter 1, I described a long cable connecting a newly rear-mounted battery with a starter motor. I said that when the car was being cranked, there was a voltage drop of 2.5V across this cable. To measure that voltage drop, you could firstly measure the voltage at the battery, and then secondly measure the voltage at the starter motor. By subtracting one from the other you would gain the voltage drop. Or instead, you could apply the negative probe of the multimeter to the power cable at the battery, and the positive probe of the multimeter to the cable at the starter motor, and then read the voltage drop off directly as the engine cranked.

Measuring voltages probably makes up 90 per cent of uses for a multimeter on a car.

USING A MULTIMETER TO MEASURE OHMS

Resistance is measured differently to voltage. Any component having its resistance measured *must not be in a complete circuit*. This means it must be disconnected at least at one end (or usually, in practice, at both ends).

For example, if you want to measure the resistance of a coolant temperature sensor, the sensor should be unplugged from the wiring loom, and then the multimeter probes connected to the two sensor terminals. The meter is then set to measure resistance, and the reading is taken. Back-probing (as we did above when measuring voltages) should typically *not* be done when measuring resistances.

In addition to temperature sensors, other common resistance measurements on cars are of injectors, relay coils, solenoid coils, and potentiometers (eg unplugged throttle position sensors and suspension height sensors).

3. MULTIMETERS

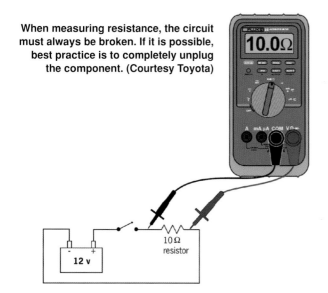

When measuring resistance, the circuit must always be broken. If it is possible, best practice is to completely unplug the component. (Courtesy Toyota)

Measuring the continuity of a coolant temperature switch. This one is open-circuit at room temperature (which is what you'd expect!).

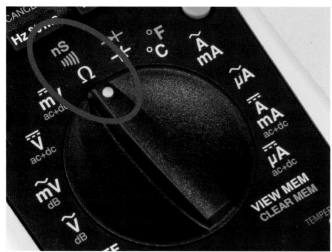

Setting a multimeter to read resistance (Ω). Note that in this meter a press of another button gives the continuity function, where a beeper sounds when there is a complete circuit between the probes.

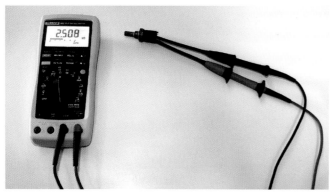

Measuring the resistance of a coolant temperature sensor.

USING A MULTIMETER TO MEASURE CONTINUITY

A continuity measurement with a multimeter simply confirms that the circuit between the probes is complete. In fact, if you think about it, you could use the resistance measurement function to do this. However, on a multimeter, when continuity is complete, the meter usually sounds a beeper. This makes practical measurements quicker and more certain – if the beeper sounds, the circuit is complete (without any further scrutiny of the reading needing to be made).

In practice, continuity measurements are made frequently. You can quickly assess if fuses are good, if filament bulbs are not blown, and, if running new wiring, whether the wire you think is the right one, actually is. Switches and relays can also easily have their connections proved using the continuity function.

When using the function, always initially join the multimeter probes together to ensure the beeper sounds.

USING A MULTIMETER TO MEASURE AMPS

Measuring current with a multimeter is a different ball game to anything so far described. Firstly, when measuring current, the meter is inserted into the circuit in *series*, so that the current flows through the meter. Secondly, you will need to swap the positive lead of the meter to a new socket and then select the correct range of amps on the dial.

For example, if you wished to measure the current flowing to a headlight, you would need to break the circuit to the light and insert the meter. Often, pulling a fuse and then probing to each side of the fuse socket is the easiest way of doing this. You would then need to select the correct current range – always start with the highest range first. If the reading is then very small, disconnect from the circuit, select the next lowest current measuring range, and then re-apply the probes.

33

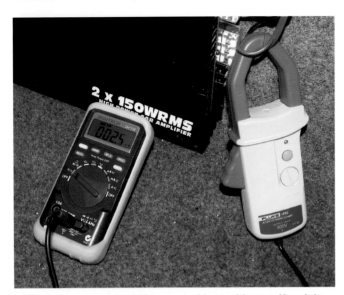

A clamp-on ammeter being used with a multimeter. Here it is measuring the current draw of an amplifier running at low levels – the displayed current is 2.5A.

When measuring duty cycle, the meter is set to volts DC, and then an additional button pressed for duty cycle, shown here as '%.'

Note that most multimeters are limited in maximum current to a short-term 10A – that's around 140W when the car is running. If you wish to measure higher currents, use a current clamp accessory (as covered previously).

Always be very careful when measuring current with a multimeter. Ensure that you initially move the probe connections as required when measuring current, then when you have finished measuring current, return the positive lead of the multimeter back to its general-purpose socket position and de-select amps on the meter dial. If you don't do these latter things, next time you try to use the meter to measure voltage, you will blow a fuse in the meter or damage it. I've been using multimeters for many years, and this is still an error I occasionally make ... and multimeter fuses are expensive!

USING A MULTIMETER TO MEASURE FREQUENCY AND DUTY CYCLE

When using a multimeter to measure either frequency and duty cycle, set the meter to volts DC, and apply the probes in the same way as you would if measuring voltage. Then *additionally* select the function you wish to measure. For example, my Fluke multimeter has a 'Hz % ms' button which, on repeated presses, selects either frequency, duty cycle or milliseconds pulse width.

If you are measuring duty cycle, you'll have a further button to press that tells the meter whether it is the 'on' times, or the 'off' times, that comprise the duty cycle you wish to measure. For example, as described earlier, a fuel injector at idle speed might have a duty cycle of only 2 per cent. However, initially the meter might show a reading of 98 per cent – an obviously absurd reading at idle. Pressing the 'invert' button will change this reading to 2 per cent.

When measuring an unknown signal with a multimeter, be careful – especially if it could be a pulsed signal. You won't do any damage, but you could get completely the wrong idea of what you are measuring. For example, if you are measuring the signal output of a speed sensor while travelling down the road, the measured DC voltage might read 2.5V. Even if you go faster or slower, that measured reading might not change – giving the impression that the sensor is broken. But if you select *frequency* measurement on the meter, the frequency might be 100Hz at low speed and 200Hz when travelling twice as fast – showing that in fact, the sensor is working. So what is going on? If the signal is a square wave with a 50 per cent duty cycle, you would simply be reading half of the max amplitude voltage of 5V (so 2.5V), irrespective of the change in frequency. This is one reason that an oscilloscope (covered in Chapter 7) is so important as a measuring tool.

When diagnosing engine management systems, measuring frequency and duty cycle are very important. Remember that frequency is a measurement of *how many times per second* something is changing, and duty cycle is the *proportion of time* something is switched on or off.

3. MULTIMETERS

LOGGING MULTIMETERS

Some multimeters are able to log the measured values. Depending on the meter, the data may be able to be exported to a PC or smartphone, or played back or graphed on the multimeter screen itself. Many people suggest that such a meter is not worthwhile – after all, you can buy data logging adaptors that will allow you to log straight to a laptop PC, and these adaptors are often cheaper than a good logging multimeter.

However, the problem with most PC data loggers is that they are strictly limited in their input voltage, and don't have the 'circuit smarts' of a multimeter. For example, a PC logger will typically have a maximum input voltage of 5V, meaning that voltage divider circuits (see Chapter 6) will need to be used to drop higher voltages to levels that are safe for the unit. Furthermore, measuring AC voltages, current, frequency and duty cycle will all be easy with a multimeter, but (although possible) much harder with a data logger.

Despite the fact that many logging multimeters display a graph of the measured value, don't confuse this sort of meter with an oscilloscope (see Chapter 7). An oscilloscope can graph very rapid changes in signals (signals that can change thousands of times a second) whereas a logging multimeter will normally sample much more slowly – eg at a maximum rate of 1Hz (once per second). Some oscilloscopes can perform logging, but not usually over long periods.

So when would you use a logging multimeter as opposed to a normal multimeter or an oscilloscope? I use a Fluke 287 – here's a case I recently made use of it.

One of my cars uses custom air suspension. At the time of writing, the system has a small leak and so I like to leave the suspension active all the time, allowing the control system to maintain the correct ride height. To achieve this, the compressor needs to operate occasionally, and the compressor draws a substantial short-term load from the battery. To replace this energy, I have fitted the car with a solar panel that charges the battery.

I left the Fluke logging battery voltage for 24 hours, and the photo on this page shows the logged voltage trace. The three dips show when the air compressor worked. The little discrete bumps are the air suspension solenoids kicking in and out. The rise in voltage is as the fog cleared in the morning and the solar panel started to work. The solar charging is an auto system that bulk charges and then float charges – these two different voltages can be clearly seen.

A Fluke 287 logging multimeter. This meter can display a graph of logged data – very useful in some situations. (Courtesy Fluke)

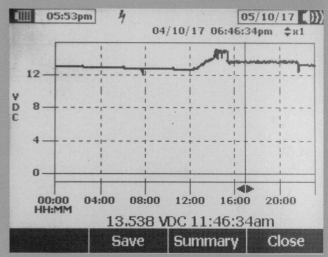

The logged voltage of a car battery. This car has air suspension that is always active, and features a solar cell to keep the battery topped-up. The three downwards spikes are the air suspension compressor running; the tiny spikes are the air suspension solenoids switching on; and the rising voltage and higher plateau are indicative of the solar cell bulk- and then float-charging the battery.

So from the logged trace I can see that:
- Battery voltage never fell very far
- The compressor ran three times in 24 hours
- The solenoids were active as they maintained the car at the correct height
- The solar controller was working properly.

And all as easily done as connecting the meter to the battery, pressing a few buttons, and walking away.

WORKSHOP PRO CAR ELECTRICAL AND ELECTRONIC SYSTEMS

Chapter 4

Fault-finding basic car electrical systems

- **Fault-finding body electrics – a 7-step approach**
- **Charging systems**
- **Starting systems**
- **Ignition systems**
- **Wiring looms**
- **Free batteries**

WORKSHOP PRO: CAR ELECTRICAL AND ELECTRONIC SYSTEMS

All cars, no matter how complex or simple, can have electrical problems. In this chapter, I want to look at fault-finding the simpler car electrical systems – ignition systems, starting systems, charging systems, lights, horns and other body electrics. I will assume that you have read and absorbed the previous chapters, and so you are familiar with different types of circuits, Ohm's Law, switches and relays, and can use a multimeter. We will start with electrics like fans and lights. Later in this book, I cover more complex computer-controlled systems – this chapter is for those working on older cars.

FAULT-FINDING BODY ELECTRICS

Car manufacturers issue manuals to train their technicians. These manuals are developed through long experience, and many start off quite simply. Despite being highly complex cars, the current Porsche technicians' manual for body electrical systems contains the following preface:

1. Do you understand how the electrical consumer is expected to operate?
2. Do you have the correct wiring diagram?
3. If the circuit contains a fuse, is the fuse OK and of the correct amperage?
4. Is there power provided to the circuit? Is the power source the correct voltage?
5. Is the ground for the circuit connected? Is the connection tight and free of resistance?
6. Is the circuit being correctly activated by the switch, relay, sensor, micro-switch, etc?
7. Are all the electrical plugs connected securely with no tension, corrosion or loose wires?

The Porsche approach is a good one, because it sequentially covers the steps that are most likely to quickly ascertain the cause of the fault. It's also applicable to nearly all basic circuits in a car. Especially if you have a good theoretical knowledge, it's easy to over-think the issue and start chasing down metaphorical rabbit holes, having overlooked the obvious and been seduced by the complex and unlikely.

So let's paraphrase the Porsche approach and ensure that if there's a body electrical problem, we can sort it out as fast as possible. Here I'll use the example of a radiator cooling fan.

1. Do you understand how the electrical consumer is expected to operate?

This is a deceptively simple question, but one that's critical to correctly answer before fault-finding. We can picture it as an imaginary conversation in our heads.

"You say the radiator fan's not working, right?"
"Yes, that's right."
"So, when is it supposed to work?"
"Er, when the engine is hot?"
"So how hot does the engine have to get before the fan is switched on? And has the engine been getting that hot?"
"Er, I think so – the temperature gauge goes up a long way and I've not heard the fan work. But, um, when I think about it, the weather has been pretty cool."

I won't continue the conversation – but you can see what I mean. If you are uncertain that the fan should in fact have been working – even with the system fully functional – then you could be chasing down a completely blind alley: there may not be any problem! On the other hand, you may be quite confident that the fan has broken – you used to hear it cycle in hot weather when stuck in traffic; now it doesn't do that and the car is overheating badly.

So before trying to diagnose a problem, be certain that the problem actually exists.

2. Do you have the correct wiring diagram?

Having the correct wiring diagram for your car is likely to make things much easier. For example, you will be able to trace the circuit and see how it is supposed to work, be able to quickly find the correct fuse and relay (if fitted), and be able to locate potentially problematic connectors.

If you do not have the correct wiring diagram, you may be able to use the owner's manual to find the correct fuses and relays, or an online search may turn up an appropriate diagram. For my cars, I always try to obtain the manufacturer's technical manuals, or failing that, a good aftermarket manual.

In this case, the fan circuit that I will be using as the example is shown in Figure 4-1. (Note that I've extracted it from a more complex diagram that also covered the air conditioner condenser fan – often you mentally need to exclude parts of the circuit that aren't of interest at the time.)

Figure 4-1 shows that the radiator fan is operated by a relay whose heavy-duty switching contact gets power via two series fuses: No 2 (80A – a fusible link) and No 11 (30A). The relay contacts feed power to the radiator fan, with the other side of the fan connected to ground. The coil side of the relay is fed power via fuses No 2 (80A) and No 1 (50A – another fusible link), the ignition switch and fuse No 16 (7.5A). The other side of the coil is grounded through a coolant temperature switch that is closed above 205°F (96°C).

3. If the circuit contains a fuse, is the fuse OK and of the correct amperage?

The statement "It's probably a blown fuse," that some people automatically make when something has stopped working, is very often correct! Let's have a look at the circuit diagram (Figure 4-1) and see what fuses we need

4. FAULT-FINDING BASIC CAR ELECTRICAL SYSTEMS

to check in this case. The two fusible links (No 1 and No 2) are very unlikely to be at fault – if either of these two fuses were blown, lots of other circuits in the car would be disabled. However, fuses 11 (30A) and 15 (7.5A) should both be checked. Note that you should always pull the fuse and check its integrity with the continuity function of a multimeter – not just do a visual inspection.

Now, if one of these fuses is blown, why did that happen?

Physically (as well as electrically), the fuse is the weakest link in the circuit. Because of this, fuses can fracture through vibration. If the fuse is the correct value for the circuit, replace it with another.

If you replace the fuse and it immediately blows, you have a short circuit to ground occurring somewhere in the system. This could be an internal short circuit in the load (if it is switched on), but it is more likely to be a wire that

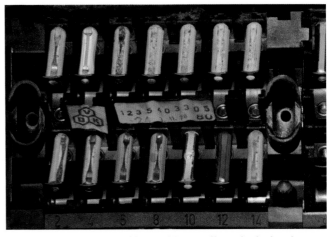

Ceramic-type fuses used in an older car. If you look closely, you can see that fuse #10 has blown.

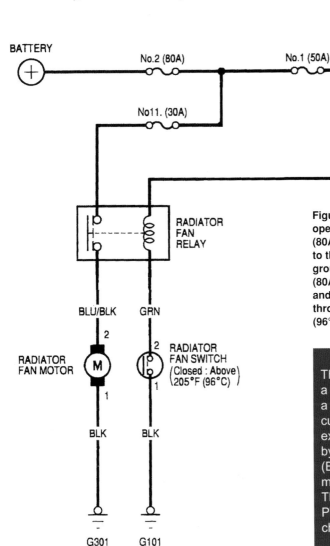

Figure 4-1: A radiator fan control circuit. The radiator fan is operated by a relay that gets power via two series fuses – No 2 (80A – a fusible link) and No 11 (30A). The relay feeds power to the radiator fan, with the other side of the fan connected to ground. The coil side of the relay is fed power via fuses No 2 (80A) and No 1 (50A – another fusible link), the ignition switch and fuse No 16 (7.5A). The other side of the coil is grounded through a coolant temperature switch that is closed above 205°F (96°C). (Courtesy Honda)

The example used here of a radiator fan is indicative of a circuit used in an older car – the cited circuit uses just a temperature switch and relay to control the fan. Some current cars use much more sophisticated systems. For example, the fan speed might be steplessly controlled by a dedicated cooling system Electronic Control Unit (ECU) that communicates by CAN bus with the engine management ECU and the Energy Management ECU. The fan might also draw over 60A at full speed and be PWM-controlled. More on these topics in subsequent chapters!

WORKSHOP PRO CAR ELECTRICAL AND ELECTRONIC SYSTEMS

Fuse blown due to short-to-ground condition

Fuse has small fracture - defective fuse

Fuse melted by an overload condition or excessive heat adjacent to the fuse

Figure 4-2: Various types of fuse failures.

This car has its fuse legend printed on the inside of the fuse box. Other cars may use just numbered fuses and need reference to the owner's manual to work out which fuse is for which circuit.

Fusible links mounted close to the battery positive. These are each 50A.

has rubbed through its insulation and is touching the metal bodywork, or something similar.

If you replace the fuse and it blows after some time (eg after the car has been running for 10 minutes), the current draw is excessive for the fuse. This could mean a low resistance to ground (again look for frayed wires and similar) but it more likely means that the device that is meant to draw current is taking a greater amount than normal. For example, the radiator fan bearings could be starting to seize, placing a larger load on the fan motor. This can easily double the current load of the fan, which over time would blow fuse No 11. In this situation, a clamp ammeter can be used to measure the actual current draw of the fan. You'd expect it to be around 75 per cent of the fuse rating (so in this case about 20A) – if it's much higher than the fuse rating, check the fan motor.

Figure 4-2 shows some different examples of fuse failures.

4. Is there power provided to the circuit? Is the power source the correct voltage?

We've checked the fuses but we also need to ensure that the wiring does not have breaks in it. The first part of doing this is to check that we have power available at the fuse – or in the case of our fan example, the fuses.

If we pull fuse No 11, we would expect to have 12V power available on one fuse terminal (the one connected to the battery) but not the other side. We can test this by grounding the negative lead of the multimeter, setting the meter to read volts DC, and then touching the positive probe to the fuse socket terminals, one at a time. After we've done fuse No 11, we can do the same for fuse No 16. We'd expect to read battery voltage on one terminal of each fuse socket – if we can't, then there is a wiring problem between that side of the fuse and the battery supply.

5. Is the ground for the circuit connected? Is the connection tight and free of resistance?

Faulty grounds are a very common cause of electrical problems, especially in older cars and those that have lived in salty environments. Most ground connections are just bolts to the metal bodywork of the car, which can become loose or corroded.

On our example fan circuit we have two ground connections – G301 for the fan motor, and G101 for the relay coil. Of the two, it's more likely to be a problem with G301 (fan motor). This is because, as I described in Chapter 1, for a given resistance, the voltage drop increases with current flow. Of the two grounds, G301 will be handling a vastly greater amount of current than G101.

The two ground connections should be inspected and wriggled. The bolts should be tight and not corroded. To

4. FAULT-FINDING BASIC CAR ELECTRICAL SYSTEMS

test the connections electrically, use the multimeter to check that there is continuity between the ground wire and the negative terminal of the battery. If you have an insulation piercing probe, you can use this to measure from the ground wire conductors; if you haven't this type of probe available, use the lug that the ground wires connect to.

6. Is the circuit being correctly activated by the switch, relay, sensor, micro-switch, etc?

There are two switches in our fan circuit – the ignition switch and radiator temperature switch. The ignition switch will not be at fault (other circuits would also stop working if it were) but the temperature switch is quite likely to be a culprit.

We can test the switch by removing it from the car and then placing it in hot water as we measure its continuity with a multimeter. The switch should close at about the nominated temperature – in this case, 205°F (96°C). However, removing the switch is likely to result in a loss of coolant – is there an easier way to test the switch? There is, and we can also use the approach to test the relay too.

With the ignition switched on, use a short length of wire to jumper the switch, that is, connect its two terminals together. The radiator fan should then run – if it does, this suggests the temperature switch is faulty and should be replaced. If the fan still doesn't run, we can listen for the relay to click when the temperature switch is jumpered. If the relay doesn't click, pull the relay from its socket and check its functionality with a power supply and a multimeter. If the relay is fine but the fan is not running, we could unplug the fan and apply 12V directly to it. If the fan doesn't run, replace the fan.

7. Are all the electrical plugs connected securely with no tension, corrosion or loose wires?

I got a little ahead of myself by testing and then replacing components in the above section – really, I should have inspected all the wiring plugs first, something I'd do at the same time as checking the ground connections. Figures 4-3 and 4-4 show some of the wiring loom issues you should be looking for.

Measuring the availability of battery voltage at a relay connection, with the relay pulled.

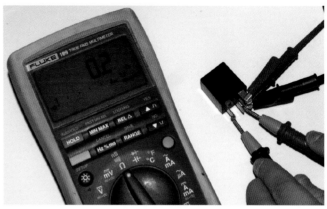

Checking a relay. The crocodile leads are supplying battery voltage to the relay coil, while continuity is being measured with the multimeter. Note the resistance reading on the display – only 0.2Ω.

Every single circuit in the car is dependent on the electrical integrity of this battery negative terminal ground connection (arrowed). Ensure it is tight and provides continuity between the battery negative post and the bodywork of the car.

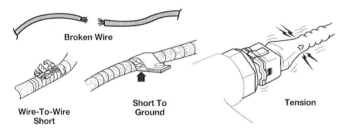

Figure 4-3: Physical inspection and testing for wiring loom plugs and sockets. (Courtesy Toyota)

WORKSHOP PRO CAR ELECTRICAL AND ELECTRONIC SYSTEMS

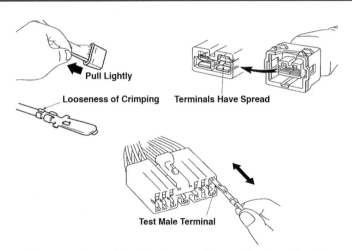

Figure 4-4: Potential wiring loom problems. (Courtesy Toyota)

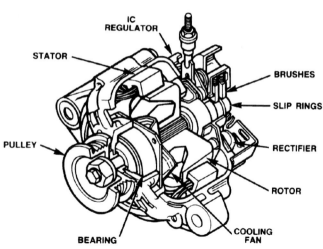

Figure 4-5: Cutaway view of an alternator. (Courtesy Toyota)

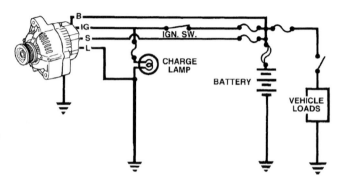

Figure 4-6: A basic alternator charging circuit. (Courtesy Toyota)

This seven step approach is one that will be successful with a wide range of body electrics. In some situations, for example a filament light not working, the first step would be to check the bulb rather than the fuse, but otherwise the approach is much the same. But what about the charging and starting systems? I'd like to look at those now.

CHARGING SYSTEMS

Nearly all cars of the last 40-odd years use a belt-driven alternator to charge a 12V lead-acid battery (the exceptions are hybrids and electric cars, which use a DC:DC converter operating off the High Voltage [HV] battery).

The alternator generates alternating current, which is rectified to produce the direct current needed to charge the battery. A voltage regulator inside the alternator controls the alternator's output to prevent battery from being over- or under-charged. It does this by regulating the flow of current from the battery to the rotor's field coil, thus changing the alternator's output.

The alternator consists of these main components:
- Stator (attached to the alternator housing – this remains stationary)
- Rotor (spins inside the stator)
- Rectifier (diodes)
- Voltage regulator

Figure 4-5 shows an alternator.

With the engine stopped, the battery provides energy to operate lighting and accessories. While the engine is starting, the battery provides energy to operate the starter motor and engine management system. When the engine is running, the alternator provides most of the required energy, with the battery then acting as a voltage stabiliser. A dashboard monitoring light (normally shown as a battery symbol) illuminates if the alternator is not providing sufficient charge. Figure 4-6 shows a basic alternator circuit.

Checking the battery

The first step in a car where the charging system is suspected of having a fault is to check the battery. Inspect the battery terminals to ensure that there is not a build-up of corrosion under the terminals, check the ground cable to ensure it has good electrical continuity with the body, and inspect the battery for low electrolyte levels and cracks in the case.

In a conventional lead acid battery, the no-load voltage is a good indicator of the battery's state of charge. The following table shows the relationship between voltage and charge. Before checking battery voltage in this way, turn on the headlights for a minute to bleed-off the surface charge.

Voltage	Per cent Charge
12.60V to 12.72V	100%
12.45V	75%
12.30V	50%
12.15V	25%

4. FAULT-FINDING BASIC CAR ELECTRICAL SYSTEMS

The voltage check described previously shows the state of charge, but it does not give an indication of the capacity of the battery. Load-testing of the battery can show a quite different story. The best electrical load for battery testing is provided by a carbon pile battery tester; this type of load can draw several hundred amps if required. In operation, it is connected straight to the battery via large spring clips.

The 'carbon pile' name comes from the internal construction: a stack of carbon discs rest against one another, with a screw knob positioned so that it can squeeze the discs together. When the discs are pressed closer together, their resistance is lower and so more current flows. When they are only loosely stacked, their resistance is higher and so less current flows. Carbon pile battery testers have large voltmeters and ammeters.

Checking a battery with one of these carbon pile testers occurs as follows:

1. Inspect the battery's rating label to identify the battery's CCA (Cold Cranking Amps) rating. The load current that can be applied to the battery is half the battery's indicated CCA rating.
2. Connect the red battery clamp to the positive terminal on the battery, and the black battery clamp to the negative terminal on the battery.
3. Rotate the battery load tester dial clockwise until the required testing current is displayed on the ammeter.
4. After 15 seconds of the load being applied, a buzzer will sound. Read the voltmeter.
5. If the battery is good, the voltmeter should display 9-10V at the time the buzzer sounded. (10V at 21°C [70°F], and 9V at -18°C [0°F]). If the voltmeter reads lower than this, the battery is defective.

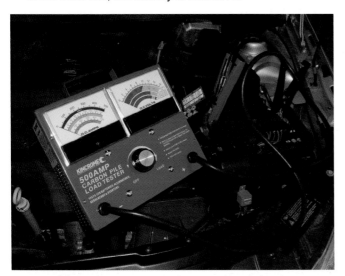

A carbon pile battery tester like this can apply large loads to the battery, testing its capacity as well as its voltage.

Checking the charging system

The following table shows the next steps to take in diagnosing a charging system problem in an older vehicle.

Symptom	Possible cause	Action
Charge warning light does not light with ignition on and engine off	1. Blown fuse 2. Defective warning lamp 3. Wiring connection loose 4. Defective relay 5. Defective regulator	1. Check charge, ignition and engine fuses 2. Replace lamp 3. Check voltage drop in circuit, tighten loose connections 4. Check relays for continuity and proper operation 5. Check alternator output
Charge warning light does not go out with engine running; battery overcharged or undercharged	1. Loose or worn drivebelt 2. Defective battery or battery connections 3. Blown fuse or fusible link 4. Defective relay, regulator or alternator 5. Defective wiring	1. Check drivebelt, adjust or replace as needed 2. Check battery and its connections 3. Check fuse and fusible link 4. Check charging system output 5. Check voltage drop
Noise	1. Loose or worn drivebelt 2. Worn alternator bearings 3. Defective diode	1. Check drivebelt, adjust or replace as needed 2. Replace the alternator 3. Replace the alternator

Measuring battery voltage when the engine is running will show you the alternator charging voltage. In the past, many alternators outputted 13.8V as the nominal charge voltage. However, over time what was regarded as the normal output gradually rose to 14.4V; some manufacturers then said that a charging voltage of 13-15V was fine. At least one manufacturer currently specifies that with their valve-regulated, lead acid, absorbed glass mat (VRLA AGM) batteries, the alternator charging voltage should be 14-16.5V! As a rule of thumb, with the engine running and no extra loads applied, the voltage measured at the battery should be around 14V.

That's charging voltage – but what about alternator output current? To measure this, you will need a clamp-style ammeter and a large electrical load like the carbon

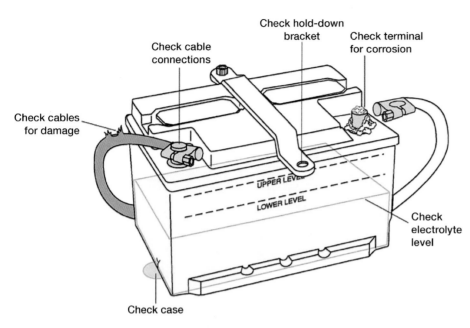

Figure 4-7: The first step in diagnosing charging problems is to carefully inspect the battery. (Courtesy Toyota)

FREE BATTERIES

It's a little-known fact that many car batteries that are discarded are still in good condition. In fact, on my sampling, I'd say that around one-third of discarded batteries are fine for at least a year or two of car use, and around half of the discards are perfectly useable for applications like solar lighting systems or solar-powered electric fences on farms.

If you're interested in scoring zero-cost batteries, here's what to do. Strike up a relationship with a local garage or battery specialist that will let you have some discarded batteries, either at zero cost or at a very low nominal cost. Using your multimeter to make the measurement, take only those batteries that still have at least 11.5V no-load voltage. Also select those where you can access the cells. This might mean older-style batteries that have refill caps on each cell, or ones where a lid or sticker needs to be lifted to reveal the cell caps. Don't take completely sealed batteries.

Take the batteries home and check the fluid levels. Where the fluid level is low (ie plates revealed) inspect the plates and if they appear unwarped, add demineralised or distilled water until the plates are covered. (If the plates are warped, you probably don't have a battery that is useable.)

Place the battery on a slow charge (say 5A) until the voltage of the battery stays at around 12.6V when it has been briefly loaded (eg via a car headlight) and then left for 30 minutes.

Using a battery tester like the one described earlier in this chapter, or an equivalent heavy load, check the battery voltage when it is loaded. For example, use a 150A load for 10 seconds. (I use a 50A load on small motorcycle type batteries.) If the voltage stays above about 10V, you almost certainly have a good battery.

More contentious is treating the battery to reduce sulphation. I've tried various 'home remedies' but the cheapest, simplest, least toxic and most readily available is Epsom Salts (magnesium sulphate). This is sold in supermarkets packaged in small flakes; people put it in the bath! Add a teaspoon of Epsom Salts to each battery cell, discharge the battery (eg by a car headlight) and the recharge. Repeat the cycle, and you should find that you can draw a greater and greater current from the battery.

An example. I started with a small 12V battery from my tractor lawnmower; I regret to say in my ownership, the battery had spent a lot of time dead-flat. Initially, using the carbon pile battery tester, I could draw only 5A before the voltage sagged disastrously. I added the Epsom Salts, and worked the battery through charge/discharge cycles for 24 hours. I then found I could draw 25A without the voltage sagging too far (to less than 10V). Some more Epsom Salts and another 24 hours of cycling, and I could draw 35A. Another 24 hours of cycling and I could draw 50A, and after yet another 24 hours of cycling, I could draw 100A. An astonishing recovery.

I have 12V lead-acid batteries that I use to power my solar garden lighting, jump-start cars in my home workshop, even use to occasionally run a 12V water pump – and they've all been obtained in this way at zero cost.

4. FAULT-FINDING BASIC CAR ELECTRICAL SYSTEMS

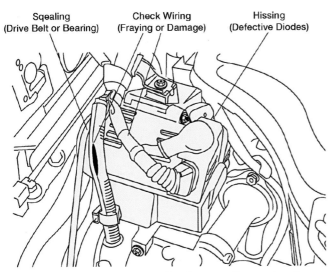

Figure 4-8: Items to check on the alternator if poor charging performance is occurring. (Courtesy Toyota)

Measuring the output of an alternator, using just car-based loads. The alternator is delivering just over 60A. Not the clamp ammeter (arrowed) around the alternator's output cable. At this current flow, the measured voltage drop between the alternator and the battery positive was just 0.1V – excellent.

pile tester described earlier. If you don't have access to such a tester, you can use the car's electrical loads instead. The largest loads comprise the lights and fans, so turn on the air-conditioning (so triggering the condenser fan), turn on the high-beam lights, and run the cabin fan at 'high.'

Connect the clamp-on ammeter to the alternator output cable (the one that goes to the battery) and zero the ammeter scale. Run the engine at 2000rpm and then apply the load. If you are using the carbon pile tester, increase the load until the battery voltage falls to 12.6V. At this voltage, the measured current should be about 75 per cent of the specified alternator output. That is, a 100A alternator should have a measured output of 75A.

The measured alternator output when you are drawing current from the battery by switching on car loads will depend not only on the health of the alternator, but how charged the battery is and how large a load you can create. On one car, I was able to measure 16A with the car running but everything switched off; 29A with the lights on, 35A when I added the cabin fan, 53A with the air-conditioning running, 61A with the high beam on, and 64A with the windscreen wipers running. That was with an alternator officially rated at 80A.

Checking voltage drops
As the charging circuit is one that carries high currents, even small resistances can cause large voltage drops. If the alternator is sensing voltage not at the battery but instead at the alternator output, a high voltage drop can lead to battery undercharging (this is because the alternator 'thinks' the battery is being fed a higher voltage than it really is). Two different voltage drop measurements should be undertaken – on the insulated (positive) side and the uninsulated (ground) side.

To make the check on the insulated side, set the multimeter to volts DC and apply the black probe to the heavy battery output terminal on the alternator. Connect the red probe of the multimeter to the positive side of the battery. Be careful on two counts: firstly, the battery terminal on the alternator is always live, so do not short-circuit this terminal to ground when you are making your connections; secondly, the engine will need to run when you are making your measurement, so ensure that none of the cables can contact moving belts or fans. The multimeter connections are often best made with crocodile clip leads, rather than the normal sharp-tipped probes. Run the engine at 2000rpm and measure the voltage drop. It should be less than 0.2V in these conditions.

You can now check the ground voltage drop. This time, connect the black probe to the alternator body and the red probe to the battery's negative terminal. Again, ensure that no leads can be caught by moving objects, run the engine at 2000rpm, and measure the voltage drop – it should also be less than 0.2V.

If you are still concerned that there may be a voltage drop under high loads, you can make the measurements again, but this time with the alternator loaded as described previously. The voltage drop will be higher, but should typically be less than 0.5V.

STARTING SYSTEMS
Nearly all cars of the last 90-odd years use an electric starter motor that turns the engine until it is rotating fast enough to start. Typically, the starter motor is equipped

WORKSHOP PRO CAR ELECTRICAL AND ELECTRONIC SYSTEMS

WIRING LOOMS

As soon as you start to work on car electrical systems, you'll be dealing with wiring looms and plugs and sockets. Plugs and sockets are used both within the loom and on various electrical devices. While these items are straightforward for experienced mechanics, for someone not used to working on cars, they can be a source of frustration and difficulty.

Wiring looms use insulated wires of different colours. Some of the colours are solid while many use additional traces. So what does this mean? A solid colour wire has just the one colour of insulation – it might be black, white, brown or red. A wire with a trace has one dominant colour and then an additional colour as a line down the wire. For example, the wire might be white with a blue trace. The trace is always the thinner of the two colours – that's important to know, because you do not want to confuse a white/blue (white with blue trace) with a blue/white (blue with white trace).

Different manufacturers use different abbreviations on their wiring diagrams for the colours; for example, black can be indicated by B, BK or BLK. These are fairly easy to understand, but what about 'L' for blue? (This is what Toyota uses.) Where a trace is used, the main colour is shown first and the trace colour shown second. For example, W-B indicates white with a black trace. Some manufacturers use idiosyncratic descriptions of the colours – for example, sky blue, violet and turquoise. I often see a colour I identify as purple, but this name is rarely used in official manuals!

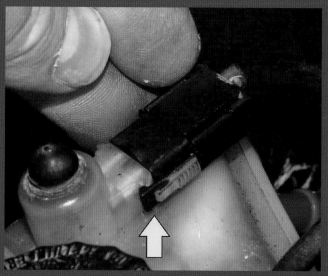

An example of a locking tab that needs to be released before the plug can be removed from its socket. There's another clip on the other side.

Especially when dealing with complex wiring systems, always double-check that your identification of the wire colour and trace are correct. If you are working on the one car, become familiar with the wiring colour codes that the manufacturer uses. If you are working on a variety of cars, ensure you know the colour abbreviations before identifying the wires.

Automotive plugs and sockets are different to those that you may have come across in other electrical equipment. They are different in three ways:

- Most have a locking tab that needs to be operated before the plug and socket can be separated. The locking tab may need to be pressed or lifted, depending on the design. Never force a plug and socket apart – if it doesn't pull apart freely, you have not released the locking device.
- Plugs and sockets located outside of the cabin have seals that make them weatherproof. Some of these seals will be inside the plug (and/or socket) and others will be located where wires enter the receptacles. When the system is apart, do not lose these seals!
- The terminals within automotive plugs and sockets can be released (normally with a specific tool) and withdrawn from the assemblies. This allows terminals to be replaced or the wiring to be changed.

Plugs and sockets in wiring looms are a major cause of electrical faults. Always treat plugs and sockets with scrupulous care when they are apart.

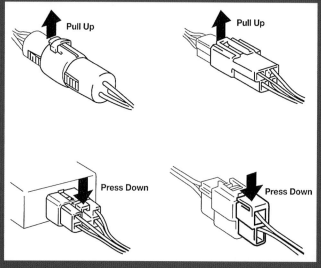

Figure 4-13: All car connectors have locks that need to be released before the plug can be removed. (Courtesy Toyota)

4. FAULT-FINDING BASIC CAR ELECTRICAL SYSTEMS

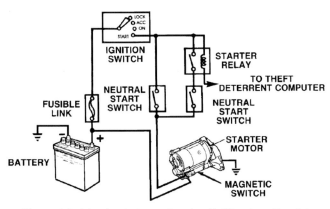

Figure 4-9: A basic starter motor circuit. (Courtesy Toyota)

Symptom	Possible cause	Action
Engine will not crank	1. Dead battery 2. Melted fusible link 3. Loose connections 4. Faulty ignition switch 5. Faulty solenoid, relay, neutral start switch, clutch switch or theft deterrent system 6. Mechanical problem in engine	1. Check battery state of charge 2. Replace fusible link 3. Clean and tighten connections 4. Check switch operation 5. Check and replace as needed 6. Check engine
Engine cranks too slowly to start	1. Weak battery 2. Loose or corroded connections 3. Faulty starter motor 4. Mechanical problems with engine or starter	1. Check battery and charge as needed 2. Clean and tighten connections 3. Test starter 4. Check engine and starter
Starter keeps running	1. Damaged pinion or ring gear 2. Faulty plunger in magnetic switch 3. Faulty ignition switch or control circuit 4. Binding ignition key	1. Check gears for wear or damage 2. Test starter pull-in and hold-in coils 3. Check switch and circuit components 4. Check key for damage
Starter spins, but engine will not crank	1. Faulty over-running clutch 2. Damaged or worn pinion gear or ring gear	1. Check over-running clutch for proper operation 2. Check gears for damage and wear
Starter does not engage/disengage properly	1. Faulty magnetic switch 2. Damaged or worn pinion gear or ring gear	1. Bench test starter 2. Check gears for damage and wear

with a magnetic switch (a solenoid) that shifts a rotating gear (the pinion) into and out of mesh with the ring gear on the engine flywheel. The solenoid is also used to switch the very high current flow from the battery, meaning that the ignition switch starting circuit needs to flow only a low amount of current.

Figure 4-9 shows the starter circuit in an older car. Note that the starter circuit may contain a neutral start switch (allowing the engine in an automatic transmission car to be started only in Park or Neutral), a clutch switch (in a car with a manual transmission, allowing the engine to be started only when the clutch is depressed), or a theft deterrent relay input (allowing the car to be started only when the car has been unlocked by the correct remote, or the correct ignition key has been inserted).

Once the engine starts, an over-running clutch prevents the engine from driving the starter motor. Some starters are direct drive while others use internal reduction gears. Figure 4-10 shows a reduction gear starter motor.

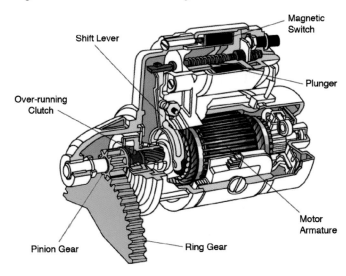

Figure 4-10: A starter motor. (Courtesy Toyota)

Checking starter cranking voltage

The starter draws more current than any other electrical device in the car. On some cars, the starter motor can draw well over 200A. This means that for a given resistance, the voltage drop will be greatest in the starter circuit.

To test for cranking voltage drop, connect the red probe of the multimeter to the positive side of the battery. Connect the negative lead to the battery terminal on the starter motor. As with alternator testing, be careful that you

WORKSHOP PRO: CAR ELECTRICAL AND ELECTRONIC SYSTEMS

Measuring the current draw of the starter motor. This four-cylinder Mercedes is drawing 151A – about what you would expect.

do not short-circuit this terminal to ground when you are making your connections, and be certain that none of the cables can contact moving belts or fans.

You will need to prevent the engine from starting during the test. With electronic fuel-injection cars, pulling the EFI fuse or relay, or the fuel pump fuse or relay, will prevent the engine from starting. In non-EFI cars, pulling the ignition system fuse or taking the high voltage lead off the distributor will do the same.

Crank the engine and read off the voltage drop value - it should be 0.5V or less. If it is higher than this, check the heavy-duty cable (and its connections) between the starter and battery positive for security and high resistances. Don't forget that the ground circuit for the starter is made through a connection between the engine and ground, or between the engine and the negative terminal of the battery – so check these too.

Checking starter cranking current draw

The current being drawn by the starter motor during cranking can indicate the presence of problems. To do this test, first ensue that battery voltage is at least 12.6V. Place a clamp-on ammeter around the battery cable that leads to the starter motor, and zero it. Again, you will need to prevent the engine from starting during the test. Crank the engine with the starter and read off the current draw.

If you have a workshop manual, you can compare the measured value with the factory specification. In general terms, a four-cylinder engine will draw about 150A, a six-cylinder will draw about 200A, and an eight-cylinder will draw about 250A. An overly high value indicates that the starter motor is faulty, or the engine is hard to crank (eg because a carbon build-up is causing higher compression). A very low value indicates a high resistance in the battery supply wiring.

Checking ignition systems

I explore engine management in detail in later chapters, but what if you're dealing with a car with a simple ignition system? A conventional points-based, distributor ignition system comprises the following components:

Component	Function
Ignition coil	Develops and stores the ignition energy and delivers it in the form of a high voltage surge through the ignition lead to the distributor
Ignition switch	Key-operated switch in the primary of the coil
Ballast resistor	Shorted (bypassed) during starting, giving a voltage boost
Points (contact breaker)	Opens and closes the primary circuit of the ignition coil, providing timing of the spark
Condenser	Provides for low-loss interruption of the primary current and suppresses arcing of the points
Distributor	At the time of ignition, distributes the firing voltage to the spark plugs in a pre-set sequence (the firing order)
Centrifugal advance mechanism	Automatically advances ignition timing with increasing engine speed
Vacuum advance mechanism	Automatically shifts the ignition timing on the basis of engine load
Sparkplug	Contains the electrodes across which the spark occurs

Figure 4-11 shows an ignition system of this type.

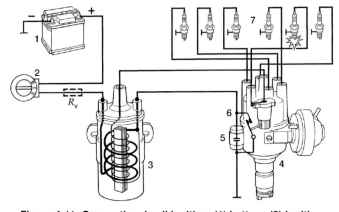

Figure 4-11: Conventional coil ignition. (1) battery, (2) ignition switch, (3) coil, (4) distributor, (5) condenser, (6) points, (7) spark plugs, (Rv) ballast resistor (not always fitted). (Courtesy Bosch)

48

4. FAULT-FINDING BASIC CAR ELECTRICAL SYSTEMS

In most cars of the last 30 years, the points are replaced by a breakerless system (eg using an inductive or Hall Effect sensor) controlling a switching transistor (an 'ignition module'). In later cars again, the spark advance is electronically controlled. In cars with engine management (covered later), the ignition and fuelling systems are integrated and are likely to use individual coils for each plug.

If the engine does not start, test for spark by removing a sparkplug and laying it on the (metal!) rocker cover. A spark should be able to be observed at the sparkplug when the engine is cranking. If there is spark and the engine does not start, the problem must be in the ignition timing or fuel delivery.

If there is no spark, inspect the points (contact breaker). Check that they open and close normally when the engine is rotating, that their contact surfaces are in good condition, and that its connecting terminal is tight. Opening the points while the ignition is switched on should result in a small spark across the contacts. If there is no such spark, use a multimeter to check that there is battery voltage at the positive terminal of the coil. If there is not, check for an open circuit in the supply wiring. If there is battery voltage available at the coil, move your attention to the high voltage side of the system. In a system with or without points, connect your multimeter (or a test light) to the negative side of the coil. The multimeter should fluctuate (or the test light blink) as the engine is cranked, indicating that the primary current is being turned on and off. If there is no pulsing on the negative side of the coil, the problem is a defective pick-up (eg the inductive or Hall Effect sensor), ignition module or wiring.

Detach the coil high voltage lead and, using a suitable insulating holder, position it about ¼in (6mm) from a ground component (eg the metal rocker cover). A spark should be seen here when the engine is turned over. If there is, the problem may be in the distributor cap – a defective rotor or a weakened spring preventing the brush making contact with the rotor. No spark indicates a fault in the high voltage winding of the coil.

If the coil is suspected of being the problem, use a multimeter to measure the primary (low voltage) coil resistance. Do this by measuring between the two low voltage terminals. The measured resistance should be in the range of 0.4-2Ω, not a short circuit or very high resistance. Also measure the resistance between the positive primary terminal and the high voltage (secondary) output. This resistance should typically be 6-15kΩ – again, not an open or closed circuit.

In electronic ignition systems, the Hall Effect or inductive sensor might also be at fault. Testing of these sensors is best done with an oscilloscope, covered later in this book.

If the engine is misfiring, run the engine at a fast idle and successively short each sparkplug to ground using an insulated screwdriver. If the engine speed does not drop when a plug is shorted in this way, that is the cylinder that is misfiring. Regular misfiring in a cylinder is usually caused by a faulty sparkplug or lead. Faulty leads can sometimes be found through a close inspection – the lead will have surface marks where arcing to ground has been occurring. Intermittent misfiring may be due to dirty or pitted points, wrong setting of the points gap, incorrect ignition timing or defects in the distributor cap, rotor or condenser. If all the cylinders are misfiring, also check the fuel system.

Spark plugs may show an indication of the problem. If the sparkplug has dry, black carbon deposits, the problem may be:
- Excessive idling
- Slow-speed driving under light loads that keeps the sparkplug temperatures too low to burn off the deposits
- Over-rich air-fuel mixtures
- Weak ignition system output

If the sparkplug has wet, oily deposits with little electrode wear, oil may be getting into the combustion chamber from the following:
- Worn or broken piston rings
- Defective or missing valve stem seals

All spark plugs should be in the same condition, and the colour of the centre insulator should be light tan or grey. If the sparkplug colour is wrong, check the heat range of the plug against the factory specifications. Highly modified engines producing more power than standard should have lower heat range ('colder') plugs fitted.

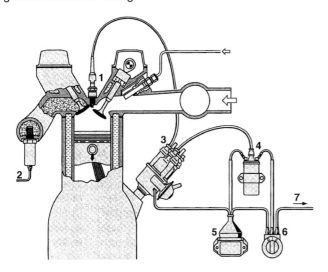

Figure 4-12: Electronic ignition. (1) spark plug, (2) oxygen sensor, (3) distributor with advance mechanisms and pulse generator, (4) coil, (5) ignition module, (6) ignition switch, (7) to battery. (Courtesy Bosch)

WORKSHOP PRO CAR ELECTRICAL AND ELECTRONIC SYSTEMS

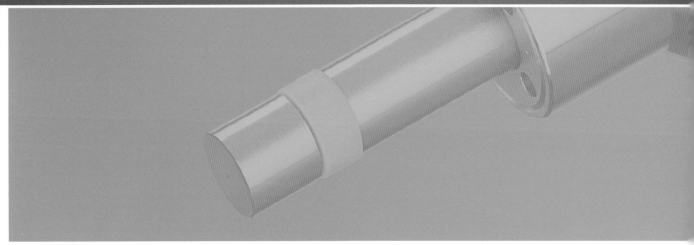

Chapter 5

Analog and digital signals

- Analog signals
- Digital signals
- Frequency and duty cycle
- Pulse width modulation
- Data bus signals
- Measuring analog and digital signals

Let's look now at the type of signals that are used within car circuits. But first, what is a 'signal?' Instead of the electricity being used to power something (for example, the lights we saw in Chapter 1), electrical signals are used to *communicate information*. This information might be the temperature of the intake air, the speed with which the car is moving, or whether the brake lights are on or off. You can think of these signals as 'coded messages' that are being sent from one place to another. The 'code' could comprise a variation in current flow but in a car, most often it is the voltage that varies. (In fact, the current flowing in automotive signal wiring is tiny.)

So how can a *variation in voltage* be used to convey information? There are two basic approaches used – these are called analog and digital.

ANALOG SIGNALS

An analog voltage signal is one that changes without steps. For example, the voltage signal output of an airflow meter is usually of this type. The output may be 1.2V at idle and 4.1V at full power, and at 'in-between' loads it can be any value between these extremes. Other examples of analog voltage signals include the output signals from most MAP sensors, yaw sensors, throttle position sensors and temperature sensors.

Most analog car sensors work in the range from about 0.5 to about 5V. Analog voltage signals can be directly read with a multimeter. The voltage may vary linearly (ie proportionally) with the parameter being measured, or it may be non-linear. These ideas are most easily shown with some examples.

An airflow meter might have an output voltage of 1.0V at a mass flow of 50 grams of air per second, an output of 2.0V at 100 grams/second flow, and an output of 3.0V at 150 grams/second. That is, the voltage output of the airflow meter rises proportionally with mass airflow. The table below shows this relationship. A doubling of airflow results in a doubling of airflow meter output voltage.

Mass airflow (grams/second)	Airflow meter output (V)
50	1
100	2
150	3
200	4

Figure 5-1 graphs this relationship which, as you can see, is a straight line.

Not all analog sensors work in this way, though. An intake air temperature sensor, for example, might have a measured voltage at the ECU of 3.1V at 20°C, 2.0V

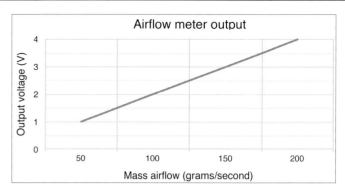

Figure 5-1: The analog voltage output of an example airflow meter. Note how the relationship between signal voltage and mass flow of air is linear and how the signal rises with increasing mass flow.

at 40°C and 0.8V at 80°C. The table below shows this relationship. Note two things – in this case, the voltage falls as the measured parameter increases, and the relationship between the parameter and the signal output is not linear.

Temperature (°C)	Sensor output (V)
20	3.1
40	2
60	1.2
80	0.8
100	0.5

Figure 5-2 graphs this relationship which, as you can see, is not a straight line.

Analog signals use two wires to transmit the information. One wire is connected to ground and the other carries the varying voltage. However, in practical applications, often you'll find three wires are used, with the extra wire feeding a regulated voltage to the sensor

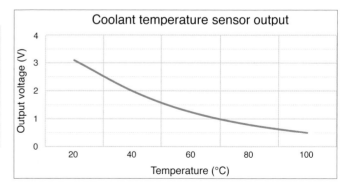

Figure 5-2: The analog voltage output of an example temperature sensor. Note how with this sensor the relationship between signal voltage and mass flow of air is not linear, and how the signal falls with increasing temperature.

5. ANALOG AND DIGITAL SIGNALS

to power it. For example, most MAP sensors have three connections:
- 5V feed
- Ground
- Signal output

Most temperature sensors are exceptions to this 'three wires' approach. These comprise variable resistors, where resistance changes with temperature. In these cases, the ECU feeds a regulated voltage to the sensor and uses an internal voltage monitoring circuit to measure the voltage change caused by the varying resistance of the sensor (more on this in the next chapter). Such sensors have only two wires – ground and signal. Figure 5-3 shows part of the circuit diagram of an engine management system. The ECU is on the left, and you can see intake air temperature (IAT) and engine coolant temperature (ECT) sensors on the right. These are analog variable resistance sensors.

Most people have no problem in understanding analog signals, so I've kept the description brief. But digital signals often cause consternation!

DIGITAL SIGNALS

Digital signals are ones that change in steps. Usually (though not always) they're either on or off. Figure 5-4 shows a graph of a digital signal varying over time. You can see that at any instant, the signal is either on (5V) or off (0V).

I said earlier that signals don't usually flow much current, but most will still light an LED – and let's use an LED as an example. If the on/off digital signal in Figure 5-4 were powering the LED, the LED would be flashing – on, off, on, off. On the other hand, with a varying analog signal voltage, the LED would instead be changing smoothly in brightness. To continue with the LED example – turning it on and off with a switch is to operate it digitally. Changing its brightness with a variable resistor is to operate it in an analog manner.

An electronic fuel injector is operated with a digital signal – the injector is either open or closed; there's no attempt made to keep it open part-way. Let's have a closer look at the way that electronic fuel injectors work, because their operation shows a lot of the characteristics of digital signals that we need to understand.

In the fuel-injection system, the fuel is supplied to injectors at high pressure. Injectors are simply solenoid valves with a built-in fine nozzle. When power is applied, the injector pintle rises, letting fuel flow through the nozzle in a spray. When power is removed, a spring shuts the nozzle, stopping the flow of fuel.

When the engine is spinning at 2000rpm, there are about 16 intake strokes every second. Since we add fuel every intake stroke, at 2000rpm we need to fire the injector (and so squirt in a bit of fuel) 16 times a second. Rather than write 'times a second,' we say the injector is being pulsed at 16 Hertz (abbreviated to Hz). This is the injector's firing *frequency*.

So if someone says that a signal varies from 50Hz to 500Hz, we know that they're referring to how many times per second the signal switches on and off – from 50 times a second to 500 times a second. As described earlier, you can directly measure the frequency of a signal using a good multimeter (the very cheapest don't have this feature).

Another example of a variable frequency signal is that used by the car speed sensor. Many older sensors work just like a bicycle speedo that has a magnet on the wheel and a reed switch on the frame. Each time the wheel rotates, it closes a contact and so sends a pulse to the ECU. If there are lots of pulses per second, the ECU knows the vehicle is travelling faster than if there are only a few pulses per second. In this case, frequency is directly proportional to speed.

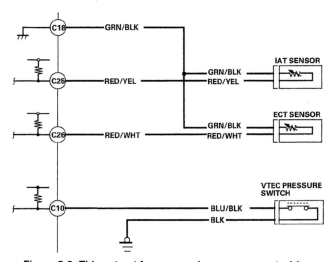

Figure 5-3: This extract from an engine management wiring diagram shows both analog and digital inputs. The Intake Air Temperature (IAT) and Engine Coolant Temperature (ECT) sensors are both two-wire, analog designs. The VTEC pressure switch, that shows when oil pressure has reached an appropriate pressure, is a digital sensor – it's either on or off. Note how the VTEC switch uses a ground return. (Courtesy Honda)

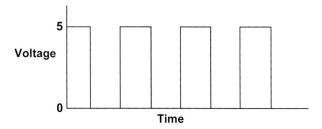

Figure 5-4: A graph of a digital signal that is either on (5V) or off (0V). Most (although not all) digital signals in cars are on/off signals

Remember that frequency refers to *how often per second the signal is changing up and down*.

Let's return to the topic of injectors. Because we want to inject fuel each intake stroke, if the engine is spinning at 2000rpm, we must open the injector 16 times a second (16Hz). Therefore, we have a maximum of 1/16th of a second to get the injector open, squirt out some fuel, and then close it again, ready for the next event. It sounds a short time, but for an injector, 1/16th of a second is long enough to take a holiday in the sun. So at these revs, it's likely it will be open for only a small proportion of the available time – say 10% of that 1/16th of a second. This percentage is called its *duty cycle*. Again, you can directly measure the duty cycle of a signal by using a good multimeter.

If the injector duty cycle is at 50%, the injector is open for half the time. If the injector is at 75% duty cycle, it is open for three-quarters of the available time. At 100% duty cycle, it is open continuously, while at 0% duty cycle it is continuously shut.

PICTURING FREQUENCY AND DUTY CYCLE

It's hard for humans to think of what's happening, all in a blink of an eye. So let's slow it right down. Instead of an injector, think of a crowd control officer at the front door of a shop. The shop is having a massive sale and the officer is opening and closing the door to let the queued-up line of people into the shop. He can't keep the door open the whole time, because then the shop would become dangerously over-crowded.

To make it fair for everyone, the crowd control officer opens the door only every minute, and to remind him, a buzzer sounds at 1-minute intervals. Each minute he decides how long to open the door. Sometimes, he opens the door for just 5 seconds out of each 60. In other words, he opens it when the buzzer sounds and closes it just 5 seconds later. He therefore has it open for 8.3% of the time (5 is 8.3% of 60).

At other times, when there are less people in the shop, he opens it for 30 seconds out of the available 60 seconds (a 50% duty cycle). When the shop is nearly empty, he opens the door for 45 seconds out of each 60 (a 75% duty cycle).

So the door is being operated with a fixed frequency (once per minute or 0.017Hz) but with a duty cycle that varies between 0 (door always shut) and 75% (door open for 45 seconds out of the available 60).

What is being described here is a *fixed frequency, variable duty cycle*. However, sometimes *both* the frequency and the duty cycle vary. For example, fuel injectors need to change in frequency, getting faster as rpm increases. We already know that they also vary in duty cycle, so with injectors, we have a *variable frequency, variable duty cycle operation*.

(A point to think about is that a 15 second door opening time at one-a-minute frequency gives a duty cycle of 25%, but that same 15 second door opening at a two-a-minute opening frequency has increased the duty cycle to 50%. That's why the measured duty cycle of injectors increases so fast with rpm – there's less time to get the fuel in.)

PULSE WIDTH MODULATION

We've seen that a digital signal can vary in two ways – its frequency and its duty cycle. This takes us to the next idea – digital control can actually be used in another way, a way that makes it act a bit like an analog signal. Let's have a look.

Most flow control solenoid valves in a car (eg turbo wastegate control valve, electronic Exhaust Gas Recirculation valve and idle speed control solenoid) used a fixed frequency but vary in duty cycle; for example, a boost control solenoid might have a fixed frequency of 30Hz and a duty cycle that varies from 10-90%. When you see that operating frequency, you might think that the flow control valve is opening and shutting 30 times per second – after all, that's what you'd expect based on what I've so far said. But in fact, when operating, valves of this sort have a variable flow that occurs without actually fully opening or closing.

Let's look at how this occurs. If the valve operating frequency is very slow (eg 5Hz), you'll be able to hear the valve actually opening and closing five times second – it will be clicking furiously. But if you speed up the valve operation a lot (and run it at say 50Hz), you can make an interesting thing happen: because of the inertia of its moving parts, at high operating frequencies the valve will hover in mid positions, with the amount of duty cycle determining how far the valve is open. In other words, the valve no longer

Fuel injectors are controlled by a variable duty cycle, variable frequency method. (Courtesy Bosch)

5. ANALOG AND DIGITAL SIGNALS

The Porsche Cayenne Turbo uses PWM control of its radiator cooling fans, allowing their speed to be steplessly controlled. (Courtesy Porsche)

completely opens and shuts at the operating frequency, but instead stays at in-between positions! The frequency that will allow this to occur is specific to the solenoid – it depends on aspects like the weight of the moving parts, the stiffness of the return spring and so on.

Of course, this variable flow approach won't work at very low duty cycles (eg 0%) because the valve will just be shut, and very high duty cycles (eg 100%) also won't work (the valve will just be fully open). Therefore, in solenoids of this sort, the actual duty cycle variation is typically in the range of 20-80%.

The same approach can also be taken with other systems. Remember back at the beginning of this section I talked about an LED? I said that turning it on and off with a switch was to operate it digitally – the LED would be either on or off. Operating it with a variable resistor was to use analog control, and its brightness would change gradually. But, just like the solenoid was held in mid-positions, we can also change the brightness of the LED by using digital control.

What we do is set a frequency that's so fast that your eye cannot see the individual flashes and so just averages the amount of light being emitted. If we trigger the LED at (say) 500Hz, we'll find that by varying the duty cycle from 0 to 100%, we can smoothly change the apparent brightness of the LED from being fully off to being fully on. At 100 per cent duty cycle, the LED sees full voltage; at 50 per cent duty cycle, the average voltage is half battery voltage.

Exactly the same approach can be taken with DC motor speed control, for examples: a radiator fan motor, water/air intercooler pump motor, or fuel pump. Let's take the example of a radiator fan. If it's being fed by a 12V supply, at 50% duty cycle the fan motor will see an average voltage of 6 volts, at 75% duty cycle it will see an average of 9 volts, and so on. This allows us to steplessly and smoothly vary the speed of the motor. This approach is sometimes called *pulse width modulation* (PWM) or 'variable duty cycle control.'

This example is different from the ones covered so far, because it involves high current flows; in fact, radiator fan current flows can be very high indeed. But PWM can also be used with very low current flows – an ECU that generates a varying voltage control signal does so using PWM.

I stated earlier in this chapter that analog signals use two wires to transmit the information. Digital systems using variable frequency and/or duty cycle to transmit information are wired in the same way – ground and signal paths. Again, as with analog sensors, sometimes you'll see three wires being used – a voltage feed to the sensor, ground and the signal wire. Refer again to Figure 5-3. The VTEC pressure switch (that detects oil pressure) is on the right – this a digital sensor in that it has just two outputs (on or off – 100% duty cycle or 0% duty cycle). Note that it uses a ground connection to form the second conductor.

DATA BUS SIGNALS

So far, we've looked at three different ways in which signals can be coded:
- Analog
- Digital via variable duty cycle
- Digital via variable frequency.

However, in addition, the communication coding can also be via a specific digital protocol comprising a succession of '0' and '1' bits. A Controller Area Network (CAN) bus takes this approach.

In a CAN bus, the information being exchanged is referred to as a message. Any linked control unit can send or receive messages. A message might contain a physical value such as the engine speed, represented as a string of ones and zeroes. For example, an engine speed of 1800rpm is represented as 00010101 in binary notation.

A CAN bus uses two wires to send signals – these wires are designated 'CAN high' and 'CAN low.' The binary

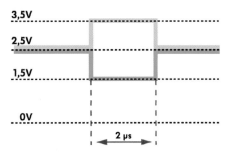

Figure 5-5: A CAN bus digital signal. Rather than being on/off, this type of signal works by driving voltages in two different directions on the two wires that form the bus. CAN low (green) goes lower in voltage at the same time as CAN high (yellow) goes higher in voltage. (Courtesy Volkswagen)

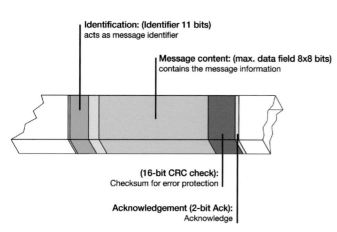

Figure 5-6: The make-up of a CAN message. Each bit is transmitted by the voltage change shown in Figure 5-5. (Courtesy Volkswagen)

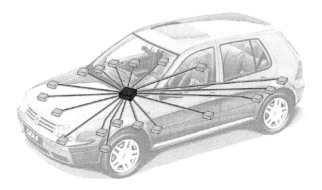

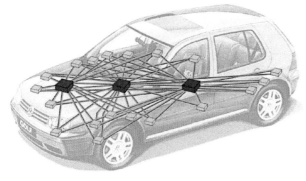

information is encoded via sudden variations in voltage between the wires. When a '0' (termed a dominant) is sent, the CAN high line is driven towards 5V, and at the same time, the CAN low line is driven towards 0V. When a '1' (termed a recessive) is sent, neither wire is driven. Figure 5-5 shows this voltage variation, with the middle of the diagram showing a '0' (dominant) being sent.

The succession of '0' and '1' signals are sent in a strictly defined format – see Figure 5-6. Data transmitted on the bus is generally received and evaluated by all the connected ECUs.

Earlier, we saw that with both analog and digital signals, devices like sensors are wired straight to the ECU, and that one side of the circuit is also grounded. However, data bus wiring is different.

First, rather than connecting sensors with control units, bus wiring always *connects control units* – it allows controllers (including dashboards) to exchange information. This approach has major advantages in terms of cost, weight and wiring complexity. By exchanging information in this way, multiple ECUs can make use of the information derived from one sensor. For example, the engine management ECU, the ABS ECU, the Electronic Stability Control ECU (and many others!) can all use the information on vehicle speed gained from one sensor. A single ECU collects this information, processes it, and then via the bus can send it to the other ECUs. An ECU can also use the input of multiple sensors to calculate something (eg 'is the cornering load high?') and then send this new information to other control units (like the Electronic Stability Control ECU and the Engine Management ECU).

Second, both the bus wiring itself, and the way it is organised, are different to conventional signal wiring. Bus wiring is made up of a 'twisted pair,' two conductors twisted around one another with neither connected to ground. Twisting them together reduces received and radiated interference. Bus systems can be configured as stars (all data lines are connected to a single point), as rings (each control unit reads the message and passes it on) or as a linear bus (the control units are each connected via short links to a 'main highway' bus).

Figure 5-7: This shows a major reason that manufacturers have adopted bus systems. The top image shows a system with one ECU controller, as might be found in a car with just engine management. The second image shows a car with multiple, separate ECUs – for example, engine management, ABS and Electronic Stability Control. Because in this approach, each ECU needs to connect to shared sensors, the complexity of wiring becomes great. The third image shows a car with a bus system, where sensor inputs (and control strategies) can be communicated between the ECUs via the bus, shown in orange. Note the reduction in wiring in the last version. (Courtesy Volkswagen)

5. ANALOG AND DIGITAL SIGNALS

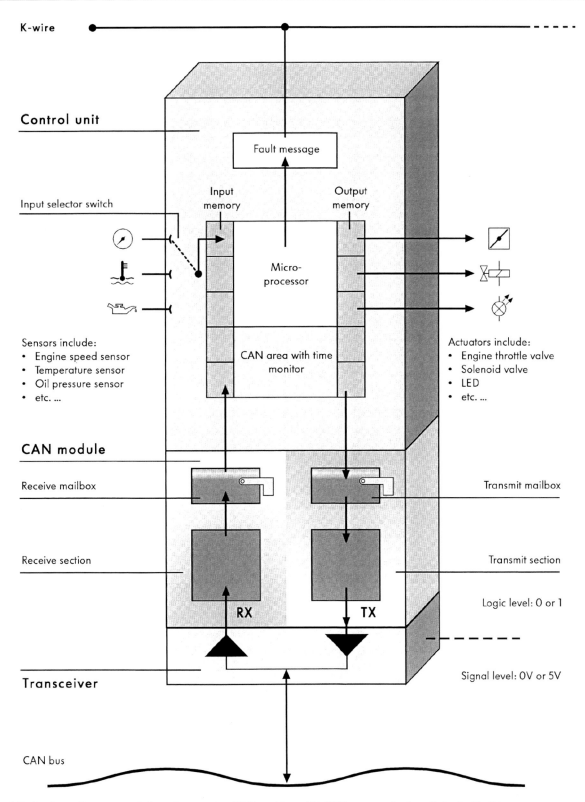

Figure 5-8: A schematic representation of an engine ECU working with CAN communication. Note that in this car, errors are reported on a K-line serial bus. (Courtesy Volkswagen)

The table below summarises the advantages and disadvantages of the different approaches.

	Advantages	Disadvantages
Star	+ Simple set-up + Relatively reliable because if one unit or circuit fails, the others can still communicate.	- If the central connection point fails, communication is no longer possible over the complete network.
Ring	+ Simple set-up + To add extra controllers, they can be added between two existing control units.	- The entire network fails in the event of an open circuit - If one control unit is faulty, the messages can become corrupted and so the network unstable.
Linear	+ If one control unit fails, the others can still communicate with each other.	- If there is an open circuit where a controller connects to the central line, several control units can fail.

In a linear bus, terminating resistors are used at each end of the 'main highway.' These prevent transmitted data from being reflected from the ends as echoes, causing data corruption. Note that there are low- and high-speed CAN buses, and other types of buses are also used (eg Local Interconnect Network – LIN, and Media Oriented Systems Transport – MOST). Figure 5-9 shows the topology of a Porsche 981 Boxster network that uses CAN, LIN and MOST buses.

MEASURING ANALOG AND DIGITAL SIGNALS

When fault-finding or making car modifications, you need to be able to measure both analog and digital signals. Sometimes, as in the case of a CAN bus for example, you might simply want to ensure that communication signals are being passed from one module to another. (CAN messages can be intercepted and changed, but that is out of scope for this book.) In other cases, such as with a throttle position sensor, you might want to check the output signal voltage against a workshop manual specification.

Measuring analog signals

As mentioned earlier, analog signals can be directly measured with a multimeter. Measuring analog signals is straightforward – you cannot damage any sensor by measuring its output with a digital multimeter. About the only thing to be wary of is when signals are changing rapidly. For example, the output of a narrow band oxygen sensor is a rapidly changing analog signal – in this case, many multimeters won't be able to 'keep up' and so you may not get a true reading. The more expensive the meter, usually the better it is at dealing with rapidly changing analog signals.

Measuring digital signals

Measuring digital signals can be much more difficult than measuring analog signals. This is because if you are not sure what sort of signal is present, it is easy to be deceived by the multimeter's reading.

Let's have a look at this idea in more detail. As I described earlier, and as shown previously in Figure 5-4, a digital signal is usually either on or off. When it is 'on,' it might have a DC voltage of 5V, when it is 'off,' it has a voltage of zero. Since it is a varying voltage, we initially set the multimeter to read 'volts DC.' If we want to measure the signal's frequency, a 'Hz' button is then pressed on the meter. If we want to measure duty cycle, a '%' button is pressed. (Depending on the model of multimeter, there may be some minor variations in how the meter settings are configured).

That all sounds straightforward – and often it is. For example, measuring injector duty cycle is just a case of:
- Connect the negative lead of the multimeter to ground (eg the negative terminal of the battery, or an electrically conductive part of the chassis).
- Set the meter to read 'volts DC,' then '%.' Connect the positive probe of the multimeter to one side of the injector (eg by pushing a sharp probe through the insulation, or by back-probing the plug).
- Drive the car and monitor the meter reading (a two-person job!).
- If there is no reading, connect the positive probe of the meter to the other side of the injector.
- If the meter's reading seems reversed (eg it shows 99% duty cycle at idle but this is not possible – it should be in this case 1%), press the 'invert' button on the meter.

So where's the problem with measuring digital signals? Okay, consider this: what if you are trying to find a signal on a wire, *but you don't know whether it's analog or digital*?

A digital signal that uses a high frequency and variable duty cycle can easily be misread as a changing analog voltage. That's because the meter will average the voltage, and if the signal is operating at high frequency with (say) 75% duty cycle, a 12V feed will be read as (0.75 x 12) = 9V. At 50% duty cycle it will 'look like' a 6V analog signal. A digital signal that uses a 50% duty cycle but varies in frequency will 'look like' a fixed voltage! That

5. ANALOG AND DIGITAL SIGNALS

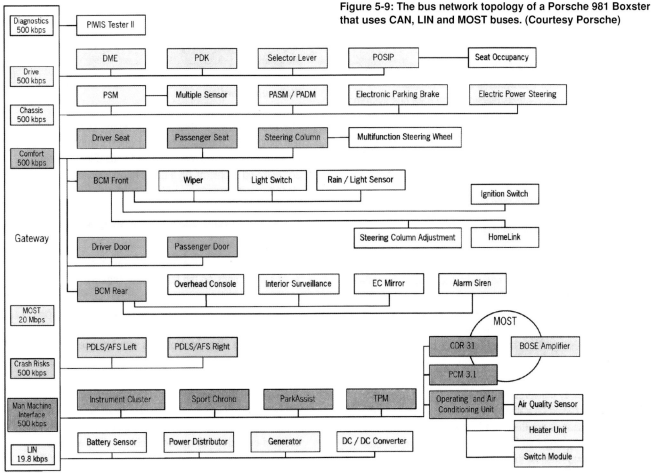

Figure 5-9: The bus network topology of a Porsche 981 Boxster that uses CAN, LIN and MOST buses. (Courtesy Porsche)

last one can be very confusing – it's easy to think: 'How come this sensor never changes its output?'

I once had this problem when trying to find the speed sensor wire behind an instrument panel. I had the front of the car jacked up (it was a front-wheel drive) and the wheels spinning slowly. I thought I'd found the right wire, but all my meter showed was a fixed voltage of 2.5V. Changing speed made no difference, and I was stumped. Then I suddenly realised it was very likely to be a variable frequency signal, and if the waveform used a 50 per cent duty cycle, I'd expect to see half the supply voltage of 5V, unvarying with speed, on my meter. And I did – it was the right wire!

You can, of course, switch the multimeter to voltage, and then frequency, and then duty cycle for every single reading, but that is laborious and unwieldy. When working with digital signals, what you really need is an instrument that 'draws a graph' of the way the signal varies over time. That way you can see if it is, in fact, a digital signal – and if it is, you can then see its frequency and duty cycle in different conditions. You can also see the shape of the signal waveform – and even if any signal is being sent at all, for example on data buses. Such a measuring instrument is called an oscilloscope. I cover oscilloscopes in detail in Chapter 7.

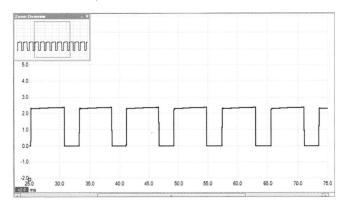

An oscilloscope screen grab, in which the scope is being used to measure the output signal of a Hall Effect sensor in an older car's distributor. A scope allows you to see the shape of the waveform as well as measuring duty cycle and frequency. More on oscilloscopes in Chapter 7. (Courtesy Pico Technology)

WORKSHOP PRO CAR ELECTRICAL AND ELECTRONIC SYSTEMS

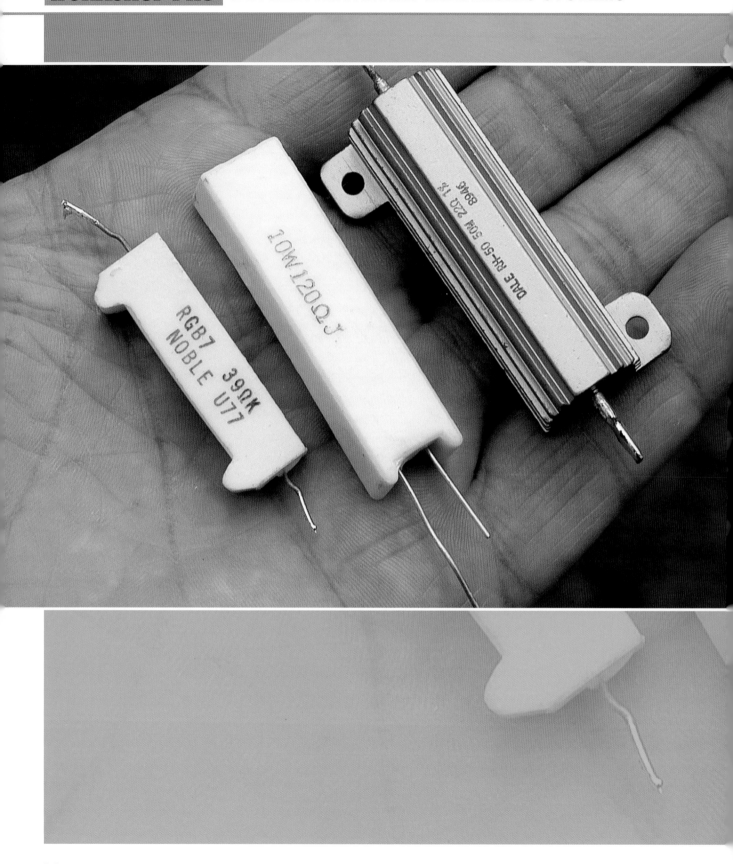

Chapter 6
Using electronic components

- **Resistors**
- **Pull-up and pull-down resistors**
- **Voltage dividers**
- **Potentiometers**
- **Capacitors**
- **Diodes**
- **Transistors**
- **Kits**

In this chapter, I want to cover some of the electronic components with which you should be familiar. The intention of this chapter is not to give you sufficient understanding that you build your own transistor or microprocessor circuits, but simply so that you can better understand car electronic systems and how they work.

A pair of 0.3Ω, 100W resistors.

RESISTORS

A resistor is an electronic component that poses a greater resistance to the flow of electricity than a normal conductor. Resistors are available in a vast range of different resistance values, from tenths of an ohm through to millions of ohms (megohms). Resistors are also rated in their ability to dissipate power, and this is expressed in watts. Small resistors use a colour code on their body to show their values, while larger resistors have their resistance written on them in ohms. Always use a multimeter to check the actual value of a resistor that you are using. Resistors are not polarised – they can be connected into a circuit either way around.

Let's look at using resistors in more detail. Let's say that you've heard how some cars vary the speed of their electric radiator fans, and you think that's a clever idea that you'd like to apply to your car. You realise that if you use a resistor in series with the fan wiring, you can drop the voltage going to the fan and therefore slow it down. If you mount a switch in parallel with the resistor, you can bypass the resistor as required, and so have a switchable two-speed fan. This circuit is shown in Figure 6-1. (Note that the 12V fed might be via a temperature switch but I haven't shown that here.)

Before we can specify an appropriate resistor to do this job, we need to know something about the fan. For example, how much current does the fan draw when connected straight to the battery, and how low do we want to drop the voltage for the slower speed? We do some measurements and find that the fan draws a steady 15A at a running-car battery voltage of 13.8V, and by working with an old flat battery, we further find that at 9V, the fan is slowed to a speed that matches what we want at the lower setting.

We therefore want to drop 4.8V (13.8V - 9V) in a circuit with 15A current flow. Ohms = volts divided by amps, and 4.8 ÷ 15 = 0.3Ω. Therefore, a 0.3Ω resistor will provide the required voltage drop. These low-ohm resistors are available, so that's fine – or is it? How much power will the resistor need to dissipate? Remember that volts x amps = watts, and we have a voltage drop of 4.8V x 15A, giving us a power dissipation of 72W. That is a lot, but 0.3Ω 100W resistors are available quite cheaply. These resistors use aluminium-finned bodies and can be bolted to a heatsink mounted in the airflow behind the grille – something I'd recommend in this case.

But note that running the fan at a slower speed in this way isn't saving any energy – rather than going into turning the fan, some of the energy is instead simply being spent on heating the resistor. Isn't there a better way? Yes: use Pulse Width Modulation (ie a varying duty cycle) as described in the previous chapter. A quick search of eBay shows a 60A, 12V PWM motor speed controller, complete with an infinitely variable control knob, for less than the price of a pizza.

One area where resistors are often successfully used to drop current flow is in the use of LEDs. LEDs can be used as dashboard indicators in cars, but connect a normal LED across 12V and the LED will immediately be destroyed. The easy way of using LEDs with 12V car systems is to place a series resistor in the circuit. The series resistor limits the current flow through the LED.

So what type of resistor do we need to allow an LED to be run off 12V? In addition to colour, intensity and package size, LEDs have two other important specs. One is what is called the 'forward voltage drop' and the other is the LED's

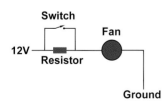

Figure 6-1: A resistor used to slow the speed of a radiator fan. The switch allows full speed to be selected by bypassing the resistor. Simple calculations can show you the resistance and power dissipation specs that the resistor needs to have.

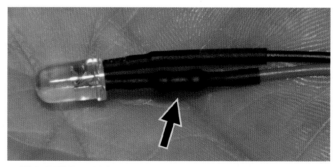

This LED is sold for use on 12V systems, and has a resistor (arrowed) already fitted under the heatshrink.

6. USING ELECTRONIC COMPONENTS

'maximum current.' With these two pieces of information (which should be available from the retailer), the required resistor can be calculated. So what type of resistor do we need to allow an LED to be run off 12V?

A bright orange LED might have these specs:
- voltage drop of 2.2V
- current of 75mA (ie 0.075A).

As we calculated for the radiator fan, ohms = voltage drop (across the resistor) divided by amps. If we are supply 12V, and we only want 2.2V at the LED, we want the resistor to drop 9.8V. So, the required resistor = 9.8V ÷ 0.075A (the required current flow through the LED) = 131Ω. Therefore a 131Ω resistor will limit the current flow to 0.075A through the LED. And the resistor's power dissipation? That's 0.075A x the voltage drop across the resistor, which is 9.8V, giving a calculated figure of 0.7W.

So we've worked out we need a resistor with 131 ohms of resistance and a power handling of 0.7 watts. The nearest off-the-shelf design to this is a spec of 120Ω and 1W, which will be fine.

You can see that if you're dealing with small currents, using resistors to create voltage drops is straightforward. We will talk about doing this in a special way (as voltage dividers) in a moment, but before that, let's discuss wiring multiple resistors into a circuit.

SERIES AND PARALLEL RESISTORS

Multiple resistors can be wired in series or parallel. Resistors placed in *series* have a total resistance in the circuit that's the same as all their individual values added up. So three 6Ω resistors wired in series have a total resistance of 18Ω. Note that they will share the power equally, so we can reduce the power rating of each by one-third. See Figure 6-2 for resistors wired in series.

3 x 6 ohm resistors

18 ohm total resistance

Figure 6-2: Resistors wired in series pose a total resistance that can be found by adding up the individual values.

And what happens if resistors are wired in *parallel*? If the resistors are all the same value, divide their individual value by the number of resistors being used, and that is the resulting total circuit resistance. So three 6Ω resistors in parallel would give a total circuit resistance of (6Ω ÷ 3 =) 2Ω, and again they would share the power dissipation. See Figure 6-3 for resistors wired in parallel.

Remember, resistors in *series* always have a total circuit resistance that is higher than any of the individual values. However, resistors in *parallel* always have a total circuit resistance that is lower than the lowest individual value.

If you have resistors of different individual values and you want to calculate the total series or parallel resistance, it's quickest and simplest to use one of the many online calculators.

3 x 6 ohm resistors

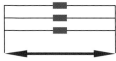

2 ohm total resistance

Figure 6-3: Resistors with the same values wired in parallel pose a total resistance that can be found by dividing the value of the resistors by how many you are using.

PULL-UP AND PULL-DOWN RESISTORS

Resistors are often used to provide signal 'pull-ups' and 'pull-downs.' But what do these terms mean?

Look at Figure 6-4. It shows a switch to 5V that is being used as an input into the ECU. An example of this sort of device might be a speed sensor. Some older speed sensors are built into the speedo of the car and comprise a reed switch that is closed whenever a magnet passes. If the magnet spins with the speedo drive, the switch will open and close rapidly as the car is driven along. The faster the car is travelling, the faster the switch is turned on and off. (Incidentally, you can see from the previous chapter that this is a digital input signal).

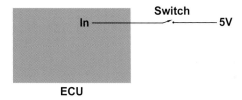

Figure 6-4: Here we have a switch input to an ECU. When the switch is closed, the input is connected to 5V, but with it open, the input is just floating – not good.

Now, let's take a closer look at what happens inside the ECU. When the reed switch is closed, there's 5 volts being fed into the ECU signal input. But what happens when the reed switch is open? Then there's nothing – the input is just floating! Any electrical noise on the input could be seen as a signal – not good.

So to avoid that problem, Figure 6-5 shows what is done. A resistor is wired between the input and ground. When the reed switch is closed, 5 volts is available on the ECU's input – the resistor is too high in value to prevent much current passing through it to ground. But when the reed switch opens, the resistor can pull the ECU input to ground. Now the input is no longer floating because it's tied to ground.

So that's a pull-down resistor – but what about a pull-up? It's very much the same idea, but this time the other side of the reed switch is connected to ground and the

WORKSHOP PRO: CAR ELECTRICAL AND ELECTRONIC SYSTEMS

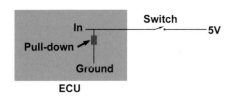

Figure 6-5: A pull-down resistor has been added. When the switch is closed, the input to the ECU sees 5V. When the switch is open, the input is pulled to ground by the resistor.

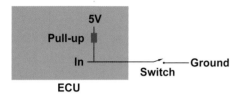

Figure 6-6: A pull-up resistor works in a similar way but the input is pulled to 5V when the switch is open, and to ground when the switch is closed.

pull-up resistor is tied to 5V. When the reed switch is open, the input to the micro is pulled-up to 5V. When the reed switch is closed, the input to the micro is pulled-down to ground. (Note the voltage of the pull-up doesn't have to 5V be – it could be 12V.) See Figure 6-6. Pull-ups and pull-downs are used with many ECU inputs.

VOLTAGE DIVIDERS

Another major use for resistors in car electronics is as voltage dividers. Figure 6-7 shows a voltage divider. Note that there are two series resistors, with 5V being fed to one end of the string and the other end being grounded. The signal output is positioned between the two resistors.

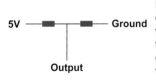

Figure 6-7: A voltage divider consists of two resistors in series, with the output taken from between them. In this case, if the values of the resistors are the same, the voltage on the output will be half the supply voltage – so 2.5V.

Let's imagine that both resistors are 10kΩ. Very little current will flow to ground (just 0.25mA) – as I said previously, current flows in signal circuits are often very small. But here's the important question: what voltage will be available on the output? In the case where the resistors are the same value, the voltage will be exactly half of the supply voltage of 5V – so 2.5V.

If the resistor on the 'ground' side is lower than the other resistor, the output voltage will be lower too. If the resistor on the 'ground' side is higher than the other resistor, the output voltage will be higher. If the value of the two resistors is known, precise calculations can be made

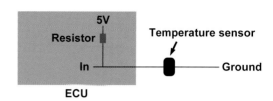

Figure 6-8: A voltage divider being used on the input from a temperature sensor. As the resistance of the temperature sensor changes, so will the voltage measured at the ECU input.

of the resulting output voltage (use any of the many online calculators if you want to do this).

Voltage dividers are used with resistive temperature sensors – Figure 6-8 shows the approach. If you think that the circuit looks very much like the pull-up resistor that we looked at a moment ago, you're right, but this time the fact that the temperature sensor is another resistor makes the circuit work as a divider. To see this, look at how the two resistors are wired in series, with the signal output taken from between the two resistors, just as we saw in Figure 6-7. The resistor inside the ECU is fixed in value, but the temperature sensor resistance changes with temperature, and so alters the value of the voltage on the ECU input.

If you look back at Figure 5-3 in the previous chapter, you'll see that the top two inputs to the ECU are voltage dividers (because the sensors are variable resistors), and the bottom input uses a pull-up. Often manufacturers will show this first component inside the ECU so that you can have a better idea of how the circuit works.

POTENTIOMETERS

Potentiometers are able to be used in two different ways – as variable resistors and as voltage dividers. It's important that you don't confuse the two. But first, what is a potentiometer, usually abbreviated to a 'pot?'

As you probably know, a pot is connected to the knob that you turn on an older radio or amplifier to make the sound louder or softer. A pot of this type consists of a semi-circular resistor track and a moveable wiper that slides along the track as the knob is turned. In the circuit diagrams that we're going to use here, the resistance track

A standard potentiometer. The two outer connections are for each end of the resistance track, and the middle connection is for the wiper.

6. USING ELECTRONIC COMPONENTS

is shown by the rectangle and the moveable wiper by the arrow. Pots have three terminals – one each end of the resistor track and another that connects to the moveable wiper.

Pots are available in different forms. They can vary in:
- Resistance – that is, the total resistance of the track
- Logarithmic or linear pots, marked A and B, respectively – you want 'B' marked pots
- Number of turns – from the normal 270° rotation right through to 15 complete turns
- Size – from tiny trim pots designed to be mounted on printed circuit boards to large wire-wound pots designed to be used as rheostats
- Design – rotary pots that turn, or linear pots that slide

When you get a pot – any pot – always check its resistance. Set the multimeter to 'resistance' (or 'ohms') and then connect the probes to the outer terminals of the pot. The meter should show the maximum value of the pot, so a 10kΩ pot should show a value close to 10kΩ (10,000Ω). It doesn't matter if it isn't exact, but if the measurement shows 4.8kΩ, for instance, then it's a 5kΩ pot not a 10kΩ pot! Next, move one of the multimeter probes to the centre terminal. Rotate the shaft of the pot and you will see the ohms read-out on the multimeter increase; if it decreases, swap the outer connection to the other side of the pot.

It is always good to test pots like this. The other day, I was doing a car modification with an unusual-design, multi-turn pot. I figured the pins were like any other pot pins and didn't check them thoroughly with the multimeter. Result? I wired the pot into the car incorrectly; and then spent a few hours chasing my tail, wondering why the mod didn't work!

A miniature potentiometer. These are normally mounted on circuit boards and are set by a screwdriver. They're very cheap, and ideal when you want to set up a circuit once and then leave it.

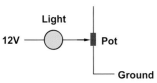

Figure 6-9: A potentiometer being used as a variable resistor. When the wiper is at the top position, the light will be dim; when the wiper is at the bottom position, the light will be bright.

When used as a *variable resistor*, only two of the pot terminals need to be used. Figure 6-9 shows such a circuit. As the pot is moved, the series resistance inserted in the circuit rises or falls, and so the light dims or brightens, respectively. Specifically, as the wiper (the arrow) is moved upwards, the interposed resistance increases, and so the light gets dimmer. The old dimmer controls used on dashboard lighting are variable resistors of this sort, sometimes called rheostats. Note that if you wish to use a pot like this, you must calculate the power that the pot will need to dissipate and spec the pot accordingly.

When used as a *voltage divider*, a pot is wired just like the voltage divider we looked at above that used fixed resistors. Figure 6-10 shows a pot wired as a voltage divider. As with the circuits we've already examined, the voltage on the output can be altered by movement of the wiper – doing this simply proportionately changes the two resistances either side of the signal output. In Figure 6-10, the voltage measurable between ground and the pot wiper will be 2.5V. As we move the wiper downwards, the voltage will fall; this is easy to remember because the wiper is getting closer to ground. As we move the wiper upwards, the voltage will rise; again easy to remember as the wiper is getting closer to the supply voltage (which in this case is 5V). Note that I have used 5V because normally a voltage divider (either a pot or fixed value resistors) will use a regulated supply, otherwise the voltage output will vary with supply voltage changes.

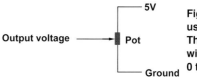

Figure 6-10: A pot being used as a voltage divider. The voltage on the output wiper can be varied from 0 to 5V

This is not a book about electronic car modification, as such, but I want to step out of the park a little and show you just how powerful the use of pots can be with car signals. First, let's take a look at using a pot to alter the input signal an engine management ECU sees.

As already described, many voltage-outputting sensors have three connections. Figure 6-11 shows the example of a manifold pressure (MAP) sensor. The top wire is the regulated 5V supply from the ECU. The centre wire is the signal output, and the bottom wire is the ground

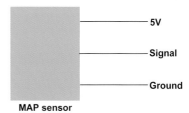

Figure 6-11: The wiring of a MAP sensor. I wanted to modify the output signal to allow me to run lager injectors.

65

connection. If you were to connect a multimeter between the ground and signal wires, you would see a voltage that varied as the engine experienced different manifold vacuums – typically a low voltage at low loads and a high voltage at higher loads.

Now let's do something rather tricky. Let's place a 10kΩ pot between the signal and 5V connections. (This might look like we're connecting the signal wire straight to 5V, but we're not – the resistance of the pot is so high that almost no current flows through this connection.) Figure 6-12 shows this circuit. Now let's measure the voltage on the pot's wiper arm – what I've labelled the new signal output. If we move the wiper up to the top, the signal output will now be a fixed 5V – and the ECU definitely won't like that! If we move the wiper to the bottom, the signal output will be exactly as it was with the standard system, so there will be no change in its behaviour.

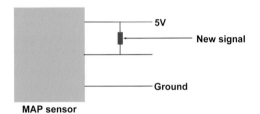

Figure 6-12: By wiring a pot between the signal and ground, we can alter the output of the sensor from being standard (wiper at bottom of track) to being fixed at 5V (wiper at top). At in-between positions, we can gain a proportion of the original signal. On a mid-Nineties turbo car, this mod allowed the running of larger injectors, retaining perfect driveability.

But what if we place the pot at an *in-between position*? Then we have a *proportion* of the original signal on the new output. So? Well I used this exact approach on a fairly simple, MAP-sensed mid-Nineties turbo car to which I'd fitted larger injectors. By adjusting the multi-turn pot, I could subtract a proportion of the load that the ECU was seeing, so returning fuelling back to standard. To put this another way, I was able to run larger injectors with an engine management modification that cost basically nothing. The car drove perfectly. Note this a modification that has extreme power over mixtures. If you choose to do it, proceed with great caution and measure air/fuel ratios at all times when you are adjusting the pot and setting-up the system.

And without getting too complex, I want you to give another example. This shows a circuit I developed to modify the electric power steering weight in a Toyota Prius. The outcome was achieved for very little money, by using just two pots.

First, let's take a quick look at the standard system – Figure 6-13 shows this. In this car, a torque sensor on

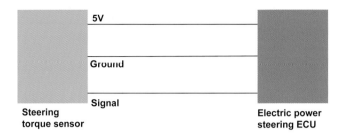

Figure 6-13: Part of an electric power steering control circuit. A torque sensor on the steering column tells the ECU how much torque the driver is inputting to the steering wheel. The output of the sensor is 2.5V with no torque, less than 2.5V with torque in one direction, and more than 2.5V with torque in the other direction. I wanted to modify the system to gain greater steering weight.

the steering column tells the power steering ECU how much effort (torque) is being input by the driver to turn the steering wheel. In its output, the torque sensor varies away from 2.5V with increasing steering torque – greater than 2.5V when the steering is turned one way, less than 2.5V when the steering is turned the other. When no driver torque is being inputted, the sensor output is constant at 2.5V. The job in this case was to *reduce the swings* away from 2.5V, so the power steering ECU thought less steering effort was occurring and so provided less steering assistance, giving firmer steering with more feel.

Figure 6-14 shows how it was done. Pot 1 is used to simply provide a 2.5V reference signal – the wiper is set half way between ground and the regulated 5V, so it will always provide 2.5V on its wiper terminal. Pot 2 allows the signal seen by the ECU to be set at anywhere from standard (pot at bottom of travel, so connected straight to sensor), to a fixed 2.5V (the voltage provided by Pot 1). So by adjusting Pot 2, the signal could be set so that it never swung from 2.5V, through to being standard in

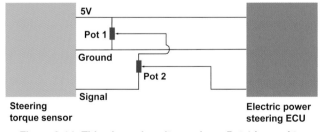

Figure 6-14: This shows how it was done. Pot 1 is used to provide a 2.5V reference signal. Pot 2 allows the signal seen by the ECU to be set at anywhere from standard (pot at bottom of travel, so connected straight to sensor), to a fixed 2.5V (the voltage provided by the other pot). So by adjusting Pot 2, the signal can be set so that it never swings from 2.5V, through to being standard in voltage swing – or any proportion in between. Therefore, by adjusting Pot 2, the amount of electric steering assistance can be set to whatever you wanted.

voltage swing, or any proportion in between. Therefore, by adjusting Pot 2, the amount of electric steering assistance could be set to whatever you wanted. As far as I know, at the time this was a world-first mod – and done just by using two cheap pots!

As I said, this book is not a manual on electronic car modification, but I've cited these examples so you can see how powerful pots can be when working with analog signals. (For more on modification, see the companion publication to this book: *Modifying the Electronics of Modern Classic Cars – the complete guide for your 1990s to 2000s car*, also published by Veloce.)

CAPACITORS

A capacitor is a device that stores a small amount of electricity. Unlike a battery, capacitors can be charged and discharged countless times without ill-effects, and this charge/discharge cycle can occur very fast. Capacitors can also be used as filters, reducing the passage of high frequencies or voltage spikes.

Electrolytic capacitors, which we'll be concentrating on here, are polarised – their negative terminal is marked with a line of negative (-) symbols. The amount of storage a capacitor can provide is shown by its rating in Farads (F). However, a Farad is a very large unit, so in typical use, you are dealing with micro-Farads, written as μF. The other rating of a capacitor is its working voltage – for example, 25V. The working voltage should always be higher than the voltage you expect to see in the circuit (otherwise, capacitors can explode!).

Capacitors can be wired in parallel to increase the value of available capacitance. Simply add their capacitance values together to reach the total, and remember to observe the working voltage requirement.

Let's first look at using a capacitor as an energy storage device. In Chapter 2 we met Solid State Relays (SSRs), devices that integrate a transistor switch into an easily-connected package. Because they use a transistor rather than a coil and contacts to do the switching, the amount of current required to operate the relay is very small. In fact, it is so small that we can use a capacitor to provide this current, giving an extended relay 'on' time.

A 10000μF, 24V electrolytic capacitor. Capacitors act as tiny storage devices of electricity and can also be used as filters.

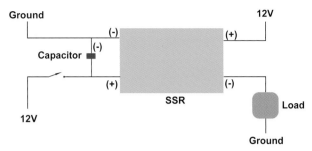

Figure 6-15: Using a capacitor to give an extended 'on' time to the output of a Solid State Relay. When the switch is closed, the relay turns on and at the same time, the capacitor is charged. When the switch is opened, the energy stored in the capacitor feeds the relay, keeping it switched on for a period. The higher the capacitance, the longer the extended 'on' time.

Figure 6-15 shows the approach. A capacitor is wired across the input terminals of the SSR, ensuring that you follow the appropriate polarities for all connections. When the switch is closed, the relay turns on and, at the same time, the capacitor is charged. When the switch is opened, the energy stored in the capacitor feeds the relay, keeping it switched on for a period. The higher the capacitance, the longer the extended 'on' time. For example, I have used a 4700μF, 25V capacitor in this way to provide an extended 'on' time of about 7 seconds. If the switch is a push-button, one press will run the item being controlled by the SSR for this period. Halve the capacitance to halve the extended 'on' time, double it to double the time, etc.

This approach can also be used as an anti-chatter mechanism. For example, if you use a sensitive vacuum switch, you can monitor when the air filter is getting sufficiently blocked to cause a restriction. Plumb the switch to the airbox after the filter, and when the restriction reaches a certain level, a simple circuit will cause a buzzer to sound. But if you do this in practice, you'll find that the switch tends to not turn on or off cleanly, as the pressure hovers around the switch-on point. A capacitor across a SSR will give a cleaner switch-on, because the system will no longer chatter – it can't, because when the SSR is switched on, it cannot turn off again until the capacitor delay has expired. Note that if you're using a large value capacitor, you can insert a series resistor into its feed, so that the charging of the capacitor occurs slowly rather than as a big gulp. This will help protect the switch from overly high current flows.

I also said that a capacitor can be used as a filter. In electronic work, the filter is usually applied at high frequencies, but it doesn't have to be. Again, an actual example best illustrates the approach. In one of my cars, I wanted to smooth the signal that the engine management ECU was seeing from the Throttle Position Sensor (TPS). This might seem to be a strange desire to have, but

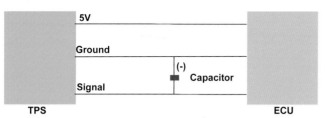

Figure 6-16: The beginning of a circuit to smooth the output of the Throttle Position Sensor (TPS). The signal output charges the capacitor, so that if the signal coming from the sensor abruptly drops, the capacitor will supply some charge, and the ECU will see a signal that is decreasing only slowly. The same smoothing will happen if the signal from the sensor rises abruptly – in that case, the capacitor will take some of this voltage as it charges and so the ECU won't initially see it.

without going into great detail, the car got much better fuel economy if it saw the driver's right foot moving only slowly. The car was a Honda Insight, and it stayed longer in a special lean cruise mode if this occurred. What was wanted was a filter that smoothed only fast changes in the TPS signal.

Figure 6-16 shows the first step. What we've done is connect a capacitor between the signal output and ground. The signal output charges the capacitor, so for example, when the signal output is 3.2V, that's the value the capacitor will be charged to. If the signal coming from the sensor abruptly drops below 3.2V, the capacitor will supply some charge, meaning that the ECU will see a signal that is dropping only slowly. The same will happen if the signal from the sensor rises abruptly – but in that case, the capacitor will take some of this voltage as it charges and so the ECU won't initially see it.

However, the circuit in Figure 6-16 has the sensor supplying current gulps to the capacitor; the sensor might not like this, and there's no way of adjusting the magnitude of the smoothing action. By adding a pot wired as a variable resistor (see Figure 6-17) we can now adjust the action of the filter and also, at all pot positions where the wiper is not at the far left, reduce the size of the current gulp the sensor needs to provide. In the Honda, I used a 10kΩ pot and a 100μF capacitor.

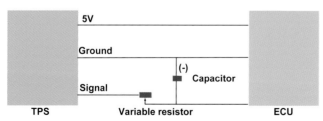

Figure 6-17: By adding the variable resistor, we now have a way of adjusting the action of the filter. The presence of the resistor also reduces the size of the switch-on current gulp taken from the sensor to charge the capacitor.

The finished TPS filter described in the main text – it uses just a pot wired as a variable resistor and a small capacitor.

Figure 6-18 shows the measured results. The screen grab is of a logging oscilloscope (more on scopes in the next chapter) and shows the voltage coming from the sensor (bottom trace) and then the voltage signal after the filter that the ECU actually sees (top trace). Note how in the highlighted area, the rapid oscillations in throttle position have been smoothed.

The Insight modification is a very unusual one that really is applicable only to that car, but it does show how a resistor and capacitor can work as a smoothing filter.

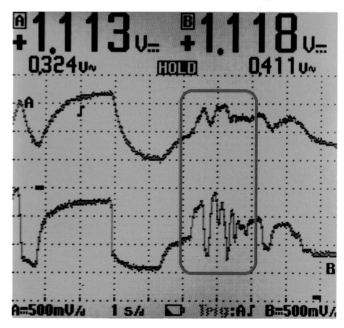

Figure 6-18: Here is a scope record of the effect of the throttle smoothing circuit. The voltage coming from the sensor is shown on the bottom trace, and the voltage signal after the filter that the ECU sees is shown by the top trace. Note how in the highlighted area, the rapid oscillations in throttle position have been smoothed.

6. USING ELECTRONIC COMPONENTS

DIODES

A diode is like a one-way valve – it lets current through in one direction only, from positive (anode) to negative (cathode). Figure 6-19 shows the circuit symbol for a diode and also, the marking that you'll find on an actual diode. These are important diagrams to look at, because they show you which way the current can flow through a diode. In the circuit symbol, *current can flow in the direction that the triangle is pointing*. Note the presence of the band on the actual diode – this represents the line that the triangle in the circuit symbol points to. When you physically have a diode in your hand, *current can flow towards the band*. A diode that is orientated to allow current to flow is called 'forward biased.'

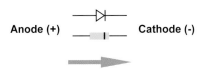

Figure 6-19: The circuit diagram symbol for a diode (top) and the appearance of an actual diode. Current can flow only in the direction that the triangle is pointing. The band on the actual component represents the line that the triangle in the symbol points to, so current can flow towards the band.

In order to turn on, a diode requires a certain voltage be applied to it – this is called the 'forward voltage.' Silicon diodes (the most common type) have a forward voltage of about 0.7V. Another way of putting this is to say that you will get a 0.7V drop across a diode. Interestingly, this is not like the voltage drop across a resistor, because the 0.7V drop doesn't vary with current flow.

When you are specifying a diode, there are two measurements that are important: the forward voltage drop, as described, and the current rating. For example, a 1N4001 diode has a voltage drop of 1.1V and a current rating of 1A. Typically, the higher the current rating, the greater the voltage drop. Most multimeters incorporate a diode check function. This allows you to determine the forward voltage drop and also the diode polarity, the later important if identifying marks have been rubbed off.

Diodes act as 'one-way valves' for current flow. They're cheap, and very useful.

Diodes are used in all car electrical and electronic systems. They are used in the alternator to rectify the Alternating Current (AC) to Direct Current (DC), they are used on body electrical systems and, as with the other components covered in this chapter, Electronic Control Units often contain many of them. But in working on a car, there are three specific uses that you're likely to make of diodes. These are for use:
* As a one-way valve
* For voltage spike protection
* To create a fixed voltage drop

Let's take a look at each of these.

One-way valve

Using diodes as one-way valves is useful when you want to link two circuits that you still wish to work largely independently. For example, let's say that you want to light a single dashboard indicator when either of two other circuits (or both) are active. Figure 6-20 shows this.

Note that we have two loads, each operated by its own switch. Load 1 turns on when the top switch is closed, and Load 2 turns on when the bottom switch is closed. But notice that closing either switch also turns on the single dashboard indicator light, that is fed from each switch through diodes. The diodes prevent the power being fed to the other load when a switch is closed.

I used a similar approach recently when setting-up an air suspension system. To deflate any of four air

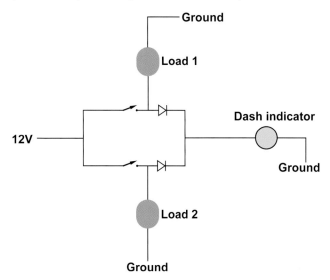

Figure 6-20: Using two diodes to allow a single dashboard indicator to light whenever either (or both) loads are switched on. Load 1 turns on when the top switch is closed, and Load 2 turns on when the bottom switch is closed. Notice that closing either switch also turns on the dashboard indicator light. The diodes prevent the power being fed to the other load when a switch is closed.

springs, individual solenoid valves needed to be triggered. However, in addition to the four solenoid valves, whenever any spring was deflating, a single exhaust valve also needed to be opened. (Think of the four deflate valves as being like the loads in Figure 6-20, and the exhaust valve as being like the dash indicator light.) The use of four diodes allowed the exhaust valve to be triggered whenever any of the four deflate valves were operating.

Voltage spike protection

Whenever a device containing a coil is switched off, a large voltage spike is generated in the wiring. This occurs with solenoids, fuel injectors and even relays. If the device is being operated by a transistor, a protection diode is often needed if the transistor isn't to suffer damage. These diodes are sometimes called 'freewheeling' diodes. With the diode fitted, the voltage spike dissipates itself freewheeling around the circuit path created by the diode. Figure 6-21 shows the way the diode is installed – ensure that you get the diode polarity correct, or you'll create a short circuit.

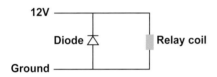

Figure 6-21: Using a diode to prevent the high voltage spike, which occurs when the relay is switched off, from damaging other components. This type of protection diode should also be used on the outputs of a SSR relay when it is controlling large inductive loads like electric motors.

Creating a fixed voltage drop

As indicated earlier, a given diode has a fixed voltage drop that varies little with current. One technique that takes advantage of this can be used to boost the charging voltage of an alternator. An alternator uses a sensing wire that detects battery voltage – when battery voltage is low, the alternator increases its charging voltage. However, some batteries are designed to have a higher charging voltage than is provided by the alternator. To boost charging voltage, simply install a diode that will drop 0.7–1V from this sensed voltage, thus causing the alternator to increase its output voltage by the same amount. In this case, the band should be closest to the alternator.

TRANSISTORS

For our purposes, it's easiest to think of a transistor as simply an electronic switch. We've seen this application in Solid State Relays, but we can also apply the idea to the internal ECU switches that control the injectors, flow control valves and mechanical relays. One of the earliest applications of transistor switches in cars was to control the primary current in the ignition coil. While some transistor outputs in ECUs are protected, many are not, so always be very careful not to short-circuit an ECU output.

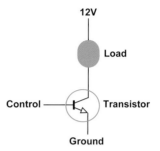

Figure 6-22: A transistor being used as a switch. When a voltage is applied on the 'control' input (it's normally through a resistor, not shown here), current can flow through the load to ground.

Figure 6-22 shows a basic transistor circuit. When a small current is applied on the 'control' input, a much larger current can flow through the load to ground. Figure 6-23 depicts how transistors are often shown in workshop manuals. This diagram shows part of the instrument panel, and you can see how transistors are used to turn on the charge, output control, and cruise indicators.

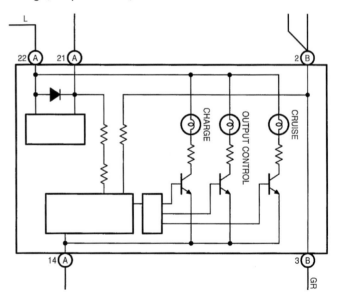

Figure 6-23: This diagram shows how transistors are often depicted in workshop manuals. In this instrument panel, transistors are used to turn on the charge, output control and cruise indicators. (Courtesy Toyota)

While in the past a book like this would have shown you in more detail how to use transistors to switch loads, these days prebuilt modules are available so cheaply that it is far easier to buy a transistor module all ready to go. In addition to the large Solid State Relays covered earlier, there are small designs available that will switch 2 to 5A; these cost less than even a conventional relay.

6. USING ELECTRONIC COMPONENTS

KITS

If you want to explore component-level electronics more deeply, one approach is to build an electronic kit. There are many kits available that can be fitted to cars – from those that add convenience and security, to those aimed at performance. Electronic kits comprise all the separate electronic components, and a Printed Circuit Board (PCB). If you're a beginner, pick a very simple kit (say, ten components) as your first kit to build.

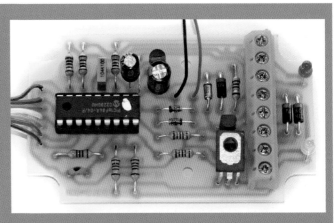

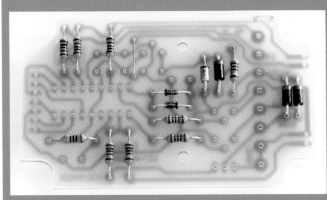

1. This is a keypad alarm kit. The first components to be placed on the PCB are always the resistors and diodes. Use a multimeter to check which resistor is which. Diodes come in different shapes and forms, but their band shows which way around they should be placed on the PCB. Select the correct component, bend its leads and then place them through the right holes. Then, turn over the PCB and solder the leads to the PCB pads. Snip off the excess lengths of wire.

3. Last to be soldered into place are the integrated circuits (ICs) and Light Emitting Diodes (LEDs), sockets and connecting leads. Here a socket has been used for the IC (sometimes, the IC is directly soldered into place). ICs must be orientated correctly to work – don't insert them the wrong way around! The polarisation of LEDs is shown by a flat on the body. No matter how strong the urge is, before you apply power, check each component. Is the orientation correct? Is it in the right place? Then turn over the PCB and check your soldering. Have you bridged any close tracks? Are any joints looking dull and suspicious, or are they all shiny-bright with the solder formed really well around the lead and track?

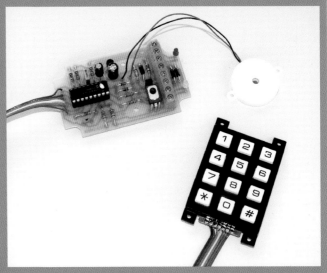

2. The transistors are the next to go on. These are three-legged polarised components, often with the legs arrange in a triangular pattern, which makes getting their orientation right a bit easier. However, some transistors have their legs all in a line, so other clues need to be used – for example, often the overlay diagram clearly specifies which way the metal back of the transistor needs to face. Next up are the capacitors. The polarised ones are cylinders marked with a line of negative (-) symbols next to one leg. Other capacitors are non-polarised (ie they can go in ether way around).

4. The alarm kit uses a remote keypad, with the keypad and the PCB connected by seven-way ribbon cable. In the original instructions, ribbon cable wasn't used – instead the two parts plugged into one another. But in this case, I wanted to mount the two parts separately, thus explaining the use of the ribbon cable. In many cases when building a kit, you can make minor changes like this – for example, you may want to use rectangular LEDs (rather than round) and mount them remotely from the PCB. A kit gives you that flexibility, but you will probably have to buy the extra components.

WORKSHOP PRO CAR ELECTRICAL AND ELECTRONIC SYSTEMS

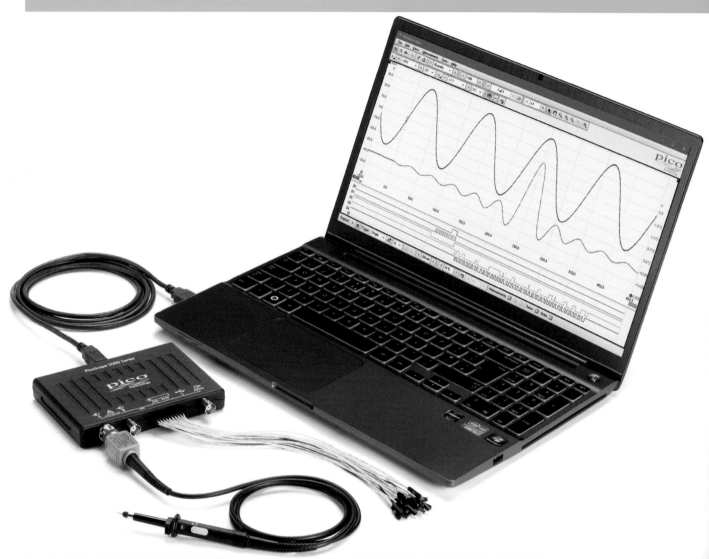

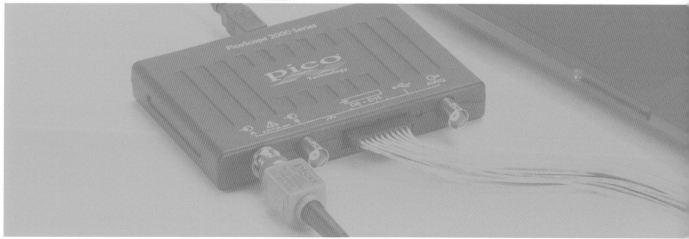

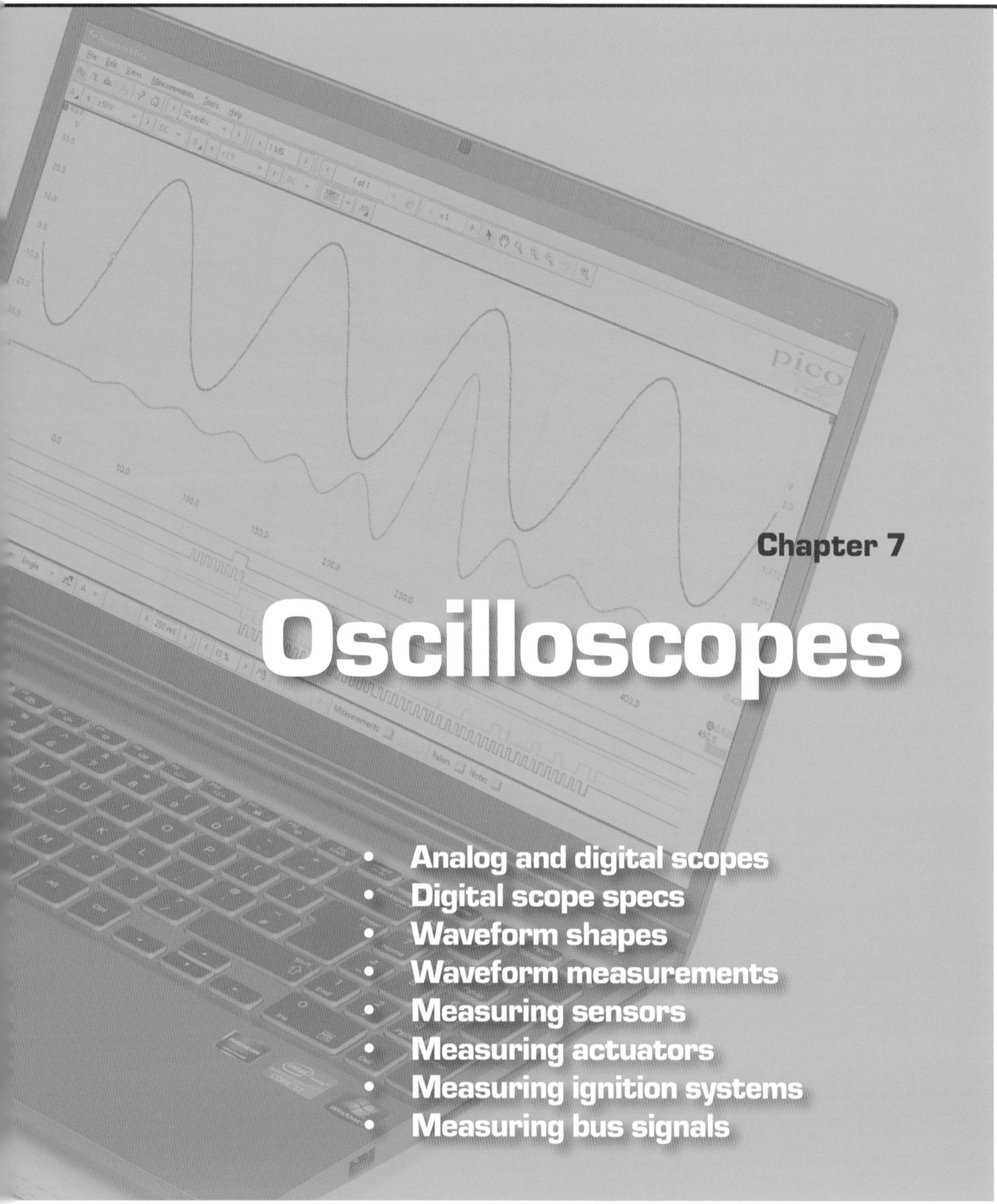

Chapter 7
Oscilloscopes

- Analog and digital scopes
- Digital scope specs
- Waveform shapes
- Waveform measurements
- Measuring sensors
- Measuring actuators
- Measuring ignition systems
- Measuring bus signals

WORKSHOP PRO CAR ELECTRICAL AND ELECTRONIC SYSTEMS

> Note: Much of the following section is based on *XYZs of Oscilloscopes* by Tektronix. Tektronix is a well-respected manufacturer of high quality digital and analog scopes.

Almost any work on car electronic systems beyond simple fault-finding of body electrics requires an oscilloscope.

An oscilloscope pictorially shows changing voltage over time, drawing a trace that accurately depicts the pattern of voltage variation. It is your window into the *shape* of the signals, whereas a conventional multimeter shows you just the magnitude of the parameter. A scope is the only way that you're going to be able to look at signals coming out of camshaft and crankshaft position sensors, speed sensors and ABS sensors, among others. And it's also the only way that you're going to be able to see the signals going to injectors, idle air control valves, boost control solenoids, auto trans pressure control solenoids, and so on.

Traditionally, scopes have been used by mechanics to look at primary (low voltage) and secondary (high voltage) ignition signals. And that's a valuable use for a scope. But these days, a scope is far more likely to be used to look at inputs and outputs of an Electronic Control Unit (ECU). In fact, many good factory workshop manuals now show sample scope traces, so that you can use a scope to quickly find out whether the output signal from the sensor or ECU looks as it should.

As suggested, an oscilloscope is basically a graph-displaying device – it draws a graph of an electrical signal. In all automotive applications, the graph shows how signals change over time: the vertical (Y) axis represents voltage, and the horizontal (X) axis represents time. This graph can tell you many things about a signal, such as:
- The time and voltage values of a signal (how many volts and when it changes)
- The frequency of an oscillating signal (how often the voltage is rising and falling)
- The frequency with which a particular portion of the signal is occurring relative to other portions (is there a part of the signal that varies more rapidly up and down than other parts?)
- Whether or not a malfunctioning component is distorting the signal (do the sine waves look more like square waves?)
- How much of the signal is noise and whether the noise is changing with time ('noise' is normally seen as a superimposed signal – jagged edges on a sine wave, for example)

I'll talk a lot more about using scopes in a moment, but first, what types are available? Oscilloscopes can be classified as analog or digital.

Analog oscilloscopes
An analog oscilloscope works by applying the measured signal voltage directly to the vertical axis of an electron beam that moves from left to right across the oscilloscope screen – usually a cathode-ray tube (CRT). The back side of the screen is treated with luminous phosphor that glows wherever the electron beam hits it. The signal voltage deflects the beam up and down proportionally as it moves horizontally across the display, tracing the waveform on the screen.

Analog oscilloscopes are characterised by the large screens used in traditional 'tune-up' machines and the smaller older scopes with the glowing green screens used in electronics work. They are excellent tools, however in automotive use they suffer from major drawbacks – the need for mains (household) power, the greater difficulty in set-up, and the absence of a storage mode that allows the freezing of the on-screen image.

Digital oscilloscopes
A digital oscilloscope uses an analog-to-digital converter (ADC) to convert the measured voltage into digital information. It acquires the waveform as a series of samples, and stores these samples until it accumulates enough samples to describe a waveform. It then re-assembles the waveform for display on the screen.

The digital approach means that the oscilloscope can display any frequency within its range with stability, brightness, and clarity. It can also easily freeze the waveform, allowing it to be studied at leisure. Digital scopes can usually be powered by batteries and use an LCD screen. All scope adaptors that are used with laptop PCs are digital. Digital scopes will usually also calculate and display the frequency and duty cycle of the signal that you are monitoring.

Three different oscilloscopes showing an interceptor's input and output signals from the crank sensor in a running car. The better the scope, the more you can see! (Courtesy ChipTorque)

7. OSCILLOSCOPES

Digital scope specifications

As briefly indicated above, an analog scope effectively draws the waveform as it occurs. However, a digital scope samples the voltages coming into the scope – it isn't continuously measuring the input signal but instead is measuring only bits of it. The waveform is then reconstructed from these separate samples and displayed on the screen. It's a join-the-dots process. How often the scope samples the signal is known as sampling speed, expressed in samples/second. All else being equal, the higher the sampling speed, the higher the frequency of the signal that can be accurately displayed. Or, to put it another way, the higher the sampling speed, the shorter the event that can be captured. In addition to sampling speed, the maximum frequency that the scope can accurately measure is also influenced by the scope's input amplifiers and filters. This factor is called 'bandwidth.'

The amount of memory that the scope has is also important, especially in automotive applications. Memory, sometimes referred to as Record Length or Buffer Size, is relevant for two reasons:
- The more closely spaced the samples are, the more memory that's required to hold them before a complete waveform can be displayed. In other words, high sampling rates require more memory.
- The longer the length of time over which the waveform needs to be displayed (called the time-base), the more samples that need to be kept if the sampling resolution is to be retained.

In automotive use, where most often quite slow time-bases are used, the second point is the more important of the two. For example, when the full width of the screen shows 90 nanoseconds, a scope may be able to sample at 10 megasamples per second, but if you lengthen the period that you want to display to 9 milliseconds, the effective sampling rate (dictated by how many samples can be memorised) may drop to only 10 kilosamples per second. In some scopes you can go even further, setting the time-base to hours! In this case you want lots and lots of memory if you're to be able to store what's basically become a data-log of the signal.

The amount of memory available is also relevant if the scope has the ability to zoom in on waveforms after they have been frozen. In order to gain that extra waveform detail, more memory will be required, especially if you have a long time-base.

In addition to sampling speed and bandwidth, the analog to digital converter (ADC) resolution of the scope is important. Most have 8-bit vertical resolution which limits the voltage variation that can be measured to just under 0.4%. On the other hand, 12-bit scopes can resolve changes in voltage levels of only 0.024%.

Especially in designs where add-on modules are used

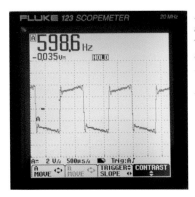

A square wave seen on an oscilloscope. Note that the scope is set to 2V/div and is using a time-base of 500μs/division. The scope measures and displays the frequency – 598.6Hz.

to turn laptop PCs into digital scopes, the functionality of the software is important. In addition to scope functions, many of these designs can also act as spectrum analysers (that is, showing on a vertical bar graph the magnitudes of all the different frequencies), multimeters (although often with quite limited ranges), and as data-loggers. All manufacturers of this type of scope allow web downloads of demo, trial or fully functioning versions of the software, so you can play before you buy the associated hardware.

Many scopes have multiple input channels, and so can simultaneously show more than one signal. This is useful when you want to see what is happening with two different signals at the same time – for example, comparing the timing of camshaft and crankshaft position sensors.

Even the cheapest single channel handheld digital scope will show you the waveform of most car input and output signals – so you'll immediately be ahead of what you'd be seeing with a multimeter. However, as you go higher in specs – especially in sampling rate and bandwidth – you can be more confident of seeing a better representation of the original waveforms.

Monitoring the output of the crankshaft and camshaft position sensors on a car running on the dyno. (Courtesy ChipTorque)

75

WORKSHOP PRO CAR ELECTRICAL AND ELECTRONIC SYSTEMS

WAVEFORMS

The generic term for a pattern that repeats over time is a 'wave' – sound waves, brain waves, ocean waves, and voltage waves are all repetitive patterns. An oscilloscope measures voltage waves. One cycle of a wave is the portion of the wave that repeats. A waveform is a graphic representation of a wave. Remember, a voltage waveform on a scope shows time on the horizontal axis and voltage on the vertical axis.

Waveform shapes reveal a great deal about a signal. Any time you see a change in the height of the waveform, you know the voltage has changed. Any time there is a flat horizontal line, you know that there is no change for that length of time. Straight, diagonal lines mean a linear change – a rise or fall of voltage at a steady rate. Sharp angles on a waveform indicate sudden change.

You can classify most waves into these types:
- Sine waves
- Square and rectangular waves
- Triangle and saw-tooth waves
- Complex waves

In automotive applications, sine and square waves dominate.

Sine waves

The sine wave is the fundamental wave shape. It has harmonious mathematical properties – it is the same sine shape you may have studied in high school trigonometry class. Mains AC voltage varies as a sine wave ('AC' signifies alternating current, although the voltage alternates too. 'DC' stands for direct current, which means a steady current and voltage, such as a car battery produces). Figure 7-1 shows a sine wave.

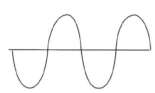

Figure 7-1: A voltage sine wave. The middle horizontal line represents 0V, so the signal is going plus and minus with respect to this. Inductive sensors output sine waves (although often not as nicely shaped as this one!)

Square and rectangular waves

The square wave is another common wave shape. A square wave is a voltage that turns on and off (ie goes high and low) at regular intervals. An injector waveform is fundamentally a square wave – the injector is either on or off. Figure 7-2 shows a square wave.

Figure 7-2: A square wave. Note that all of this wave is above the 0V line – this is not an AC waveform.

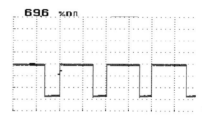

Figure 7-3: A rectangular wave, with the image taken from the scope screen. This waveform has a duty cycle of just under 70 per cent – that is, the voltage is high for 70 per cent of each repeating wave.

A rectangular wave is like the square wave, except that the high and low time intervals are not of equal length. That is, the 'on' and 'off' times are not equal. Again, this is often the case with an injector, where at low loads the 'off' time will be much longer than the 'on' time (ie variable duty cycle). Figure 7-3 shows an oscilloscope close-up of a rectangular wave.

WAVEFORM MEASUREMENTS

Many terms are used to describe the types of measurements made with an oscilloscope.

Frequency and period

If a signal repeats, it has a frequency. To remind you, frequency is measured in Hertz (Hz) and equals the number of times the signal repeats itself in one second. Hertz can also be referred to as 'cycles per second.' A repetitive signal also has a 'period' – this is the amount of time it takes the signal to complete one cycle. Period and frequency are reciprocals of each other, such that one divided by the period equals the frequency, and one divided by the frequency equals the period.

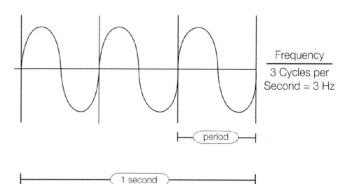

Figure 7-4: This shows a sine wave with a frequency of 3Hz – it repeats itself three times in the 1 second period. The period of the wave is therefore ⅓ of a second.

For example, a sine wave may have a frequency of 3Hz, which would give it a period of ⅓ of a second. This is shown in Figure 7-4. Some scopes can calculate frequency and display it as a standalone number, while in other cases, the period needs to be read off the scope screen and the frequency then calculated from this.

Voltage

Voltage is the amount of electric potential, or signal strength, between two points in a circuit. Usually, one of these points is ground, or zero volts. DC signals are measured on a scope from ground to the amplitude (height) of the signal. Automotive AC signals are often measured from the maximum peak to the minimum peak of a waveform, which is referred to as the peak-to-peak voltage. Figure 7-5 shows the approach to measuring peak-to-peak voltages.

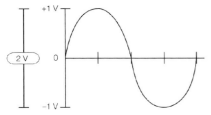

Figure 7-5: An AC waveform voltage is usually measured peak-to-peak; that is, from the bottom to the top of the wave.

USING A SCOPE

There are two major controls on a digital scope – voltage and time-base. These set the vertical and horizontal scales, respectively. Both settings are scaled per division of the screen, with screens having 8 x 10, 8 x 12, 10 x 10 or 10 x 12 graticules. Let's use a 10 x 10 screen as the example.

If you set the voltage to 2V/div, each vertical division on the graph represents 2V. That is, the display will show a maximum of (2 x 10 =) 20V. Thus, when measuring normal car electrical systems, the oscilloscope works well with the vertical scale set to 2V/div. However, if you were trying the measure the output of a narrow band oxygen sensor that has a peak output of about 900mV, using 2V per division would give you only a tiny signal on the display. In that case, 0.2V/div would be better.

Note that in most use, you simply adjust the voltage scaling until you can clearly see the signal on the screen.

The time-base sets the time per division on the horizontal axis. On the 10 x 10 graticule, a setting of 2ms per division will give you a horizontal scale that equals (10 x 2 =) 20ms, or 1/50th of a second. With most car signals that are changing rapidly (eg crankshaft and camshaft position), a setting of 20ms/div makes a good starting point. Note that CAN bus signals need a time/div setting of only around 100μs. In all cases, if you want to see more detail in the shape of the wave, reduce the time-base; if you want to see more repeating cycles, increase the time-base.

In terms of making the connections, a scope is used in largely the same way as I described in Chapter 3 for a multimeter. The earth connection of the scope (usually in the form of a crocodile clip) is connected to the car chassis and the probe is connected to the signal being monitored.

Most digital scopes have an 'auto' button that, when pressed, automatically configures the horizontal (time-base) and vertical (voltage) axes to give a good view of the signal; you can then subsequently make any voltage or time-base changes that you want. In older scopes, setting the trigger point was a major issue, but with modern scopes the software looks after nearly all of that for you.

Let's look at using a scope on a car, giving you lots of examples. It's also a good way of learning about the different signal types – eg AC sinusoidal waveforms, DC square waveforms and so on. Most of the screen grabs are courtesy of Pico Technology, a maker of good quality oscilloscope adaptors for laptop PCs.

MEASURING SENSORS

Figure 7-6 shows a scope trace from a voltage-outputting hot wire flow meter. Note how the vertical axis is in volts, starting at 0V and reaching 5V. The horizontal axis is in seconds, and covers 10 seconds from left to right (this is a very slow time-base setting). Initially, the car is idling with an airflow meter output of 0.5V. The engine is then accelerated, and the voltage very rapidly rises to about 3.6V, drops, and then climbs more slowly to peak at just under 4.5V. The throttle is closed and the airflow drops slowly as an anti-stall feature in the idle control valve stops rpm (and so airflow) dropping too fast – and what is all the noise; the spiky bits coming out of the top and bottom of the trace? The scope can react to even very short-term variations in voltage, and it is picking up some noise as well. The filter circuit in the ECU would get rid of this, and so the ECU would see only the main signal.

This scope measurement shows an airflow meter working as you'd expect. You could use a multimeter to see much the same signal shown numerically, but if there was an intermittent drop-out of the signal, or the signal was slow to rise (ie it had lag) then you wouldn't have been able to see these faults with a multimeter.

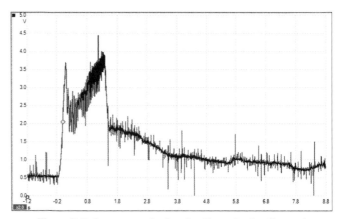

Figure 7-6: A scope grab showing the output voltage of an airflow meter as the car undergoes different loads. (Courtesy Pico Technology)

The trace we saw in Figure 7.6 shows a relatively slowly changing, analog voltage signal. Let's look at another that is much the same in type. Figure 7.7 is of the output of a coolant temperature sensor. Again the vertical axis is scaled 0-5V, but this time the horizontal axis is over 8½ minutes! This shows the voltage falling over this period as the engine warms up (the temperature increases, resistance decreases, and because it's in a voltage divider circuit, the voltage also decreases).

OK, so we've seen that we can use a scope to display a voltage signal that's changing over a period that extends from seconds to even minutes. But what about a signal that is changing fast? This is where an oscilloscope really comes into its own. Figure 7-8 shows the output of a camshaft Hall Effect sensor. This time the vertical scale extends from 0-20V, while the bottom scale is 200ms wide. To put this another way, the whole of what you can see in Figure 7-8 took one-fifth of one second to occur. The waveform is rectangular – it shows the signal at battery voltage (just over 12V) and then being abruptly pulled down to near ground (about 0.5V on this display). Each of the short downwards pulses is about 15ms wide.

Figure 7-9 shows another Hall Effect scope grab. This time the voltage is being pulled down from 5V, and you can see that in the same time span, there are more pulses. Also note that the pulses are not regular. The narrow pulse at the 60ms mark looks a little suspicious – is it meant to be there? If the scope shows that it regularly repeats, it's likely that it *is* meant to be there. But if it's erratic in its appearance (sometimes it's there, sometimes it isn't) then the sensor is faulty.

Figure 7-10 shows another camshaft sensor, but this time the scope pattern is quite different. This is an inductive sensor, a sensor that produces an output without needing a power supply. In effect it's like a small alternator, and when you look closely at the scope screen grab you can see that it is in fact an Alternating Current (AC) waveform. Note how the scale on the vertical axis has 0V half way up, and the axis as a whole extends from -5V to +5V. A waveform that is moving above and below

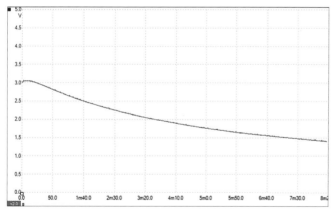

Figure 7-7: the slow change in voltage of a coolant temperature sensor as the car warms up. Note the bottom scale is about 8½ minutes. (Courtesy Pico Technology)

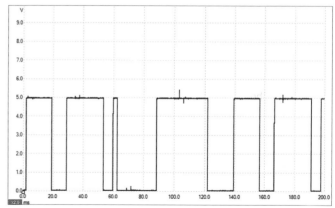

Figure 7-9: Another Hall Effect waveform. The narrow pulse at the 60ms mark looks a little suspicious. (Courtesy Pico Technology)

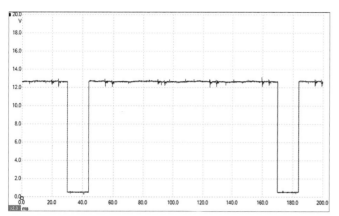

Figure 7-8: The rectangular output wave of a Hall Effect sensor on a camshaft. (Courtesy Pico Technology)

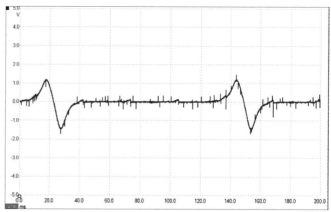

Figure 7-10: The output waveform of an inductive camshaft sensor. You can see that the waveform approximates a sine wave. (Courtesy Pico Technology)

7. OSCILLOSCOPES

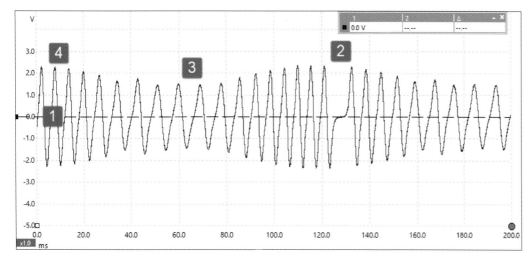

Figure 7-11: The waveform from an inductive crankshaft position sensor. (Courtesy Pico Technology)

0V is an AC waveform. The amplitude (peak-to-peak voltage) of an inductive sensor will increase with engine speed. So not only will the peaks get closer together (ie more of them displayed on the screen, even with the same time-base setting), they will also get taller. Most camshaft sensors of this type will display one output peak per 720° of crankshaft rotation.

Another sensor used to detect speed and position is the crankshaft position sensor, during cranking Figure 7-11 shows the waveform from an inductive crankshaft position sensor. (1) Indicates that the 0V axis passes through the centreline of the waveform, as it should. (2) Indicates a sudden change in the waveform caused by a missing tooth on the exciter ring. This is not a fault – the exciter wheel is made with one tooth missing so that the ECU can work out the crankshaft's rotational position. Note that the missing tooth is not necessarily at Top Dead Centre. (3) Shows the drop in amplitude with the slightly slower rotational speed of the engine as a piston is coming up on the compression stroke, and (4) shows the higher amplitude with the higher engine speed of an expansion stroke.

Figure 7-12 shows the output of both the crank position and cam sensors. The time-base has been set to bring out with clarity the cam sensor (one every 720 degrees) over six revolutions – the crankshaft pulses are running together because there are so many more of them per second. What's interesting in this screen grab is the large positive spikes reaching 12V, in addition to the expected 0-4V pulses from the camshaft position sensor. This indicates a potential intermittent short circuit to 12V.

Figure 7-13 shows another interesting scope screen grab – it's the output of a knock sensor. The sensor has been removed from the engine and is being tested on the bench by being gently tapped by a small spanner. Note how the signal output is an AC signal of about 6V peak-to-peak. The horizontal axis is 500ms wide, and the output of the knock sensor occurs for only 25ms. The knock sensor generates its own electric pulse without needing to be fed power.

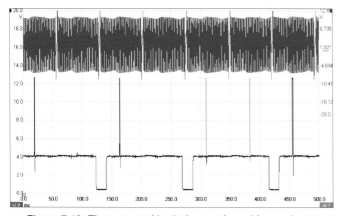

Figure 7-12: The output of both the crank position and cam sensors. (Courtesy Pico Technology)

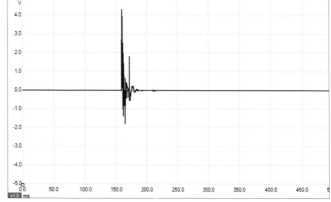

Figure 7-13: The output of a knock sensor, tested by being tapped by a small spanner on the bench. (Courtesy Pico Technology)

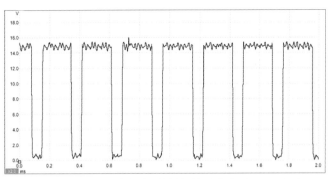

Figure 7-14: The PWM (Pulse Width Modulation) control of an EGR valve. The duty cycle is at 20 per cent - the signal is pulled to ground to turn on the valve. (Courtesy Pico Technology)

ACTUATORS

Let's look now at some actuator (output) waveforms.

Electronic Exhaust Gas Recirculation (EGR) valves are variable flow valves that are controlled by PWM (Pulse Width Modulation), a square wave with varying duty cycles. Figure 7-14 shows the pulsing of an EGR valve. The duty cycle is at 20 per cent (the signal is pulled to ground to turn on the valve). Note how the vertical axis display shows the valve is being fed from the battery. With just 2ms being the full screen width, the pulsing is occurring very fast – in fact it is shown at 3735Hz (3.735kHz). So the signal the valve is receiving is fine – if the valve isn't opening correctly, the coil is damaged or the pintle may be gummed up with carbon.

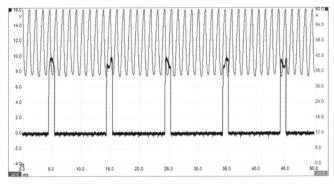

Figure 7-15: This scope screen grab shows the voltage (lower blue trace) and current (upper red trace) of a variable speed radiator fan. (Courtesy Pico Technology)

Figure 7-15 shows two traces – that is, two parameters being measured simultaneously. This scope screen grab shows the voltage (lower blue trace) and current (upper red trace) of a variable speed radiator fan. Here the fan is being operated with a duty cycle of 90 per cent (again, the signal is pulled to ground to turn the fan on). The voltage on the 'off' pulses is only about 10V, so perhaps the battery was getting a bit flat when this scope screen grab was made! The current (read from the right hand vertical axis) shows that the fan is drawing peaks of 60A.

Figure 7-16 again has voltage and current traces. This scope screen grab shows the opening of a petrol (gasoline) fuel injector. Look initially at the blue trace. Battery voltage (blue) is pulled to ground to turn on the injector; this pulse extends for just over 4ms. It then turns off and a back-EMF voltage spike of no less than 85V is emitted (don't touch working injectors!). The current trace (red) shows that the current doesn't increase anywhere near as fast as the voltage change; in fact, the inductance of the injector's coil causes only a slow current ramping-up (and so opening) of the injector.

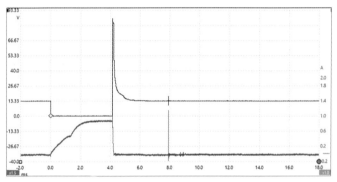

Figure 7-16: This scope screen grab shows the opening of a petrol (gasoline) fuel injector. (Courtesy Pico Technology)

The action of a diesel injector is shown in Figure 7-17. This scope screen grab shows current flow. Shown on the screen are two pre-injection squirts followed by the main injection. The injectors take a large gulp of current to switch on and then are held in that open position by a lower current created through PWM control. Using a scope on each injector in turn would quickly show if there was a wiring loom or ECU problem – the fault would show as one injector not behaving in this way.

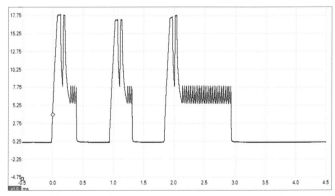

Figure 7-17: The very different scope pattern of a diesel injector, with the display showing current flow. (Courtesy Pico Technology)

7. OSCILLOSCOPES

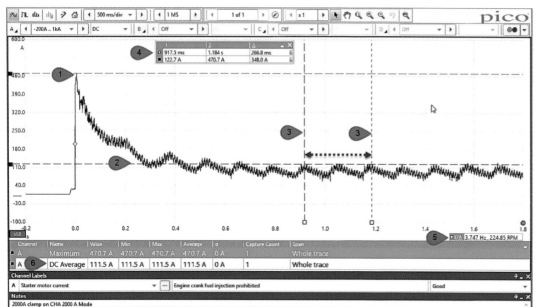

Figure 7-18: The current draw of the starter during engine cranking – this can be used to compare the compressions of the cylinders. (Courtesy Pico Technology)

RELATIVE ENGINE COMPRESSION

By using a current clamp on the starter motor feed and a scope, the relative compressions of the cylinders can be compared during engine cranking. The following description is specific to PicoScope, but similar ideas can be applied with other scopes. Figure 7-18 shows the scope screen grab.

Marker 1 shows the peak in-rush current to the starter motor required to rotate the engine from rest, which was 470A. The peak current is also recorded in the separate box shown at Marker 4.

Marker 2 shows the peak starter motor current during continuous cranking, which was 122A. The starter motor current should be even across all peaks during continual cranking as the peak current is directly proportional to the cylinder pressure during the compression stroke.

Marker 3 shows the time rulers. Align both time rulers with two consecutive compression peaks (four-cylinder engine) by clicking and dragging them. For every revolution of a four-cylinder engine, there are two compression events indicated by two starter motor current peaks during continuous cranking. PicoScope will calculate the cranking speed based on the frequency of the signal between the time rulers and record the values in the frequency/RPM legend shown at Marker 5.

Marker 6 shows the measurement table indicating peak in-rush and average starter motor cranking current.

IGNITION COILS

Scopes have been used for decades on ignition systems, and a scope remains a very useful diagnostic tool for ignition systems of all types. Note that when measuring the primary side of the ignition coil, you must use an attenuator to decrease the voltage that the oscilloscope sees – PicoScope recommends a 10:1 attenuator with its scopes. The attenuator is needed because voltages on this side of the system can reach 400V when the coil's magnetic field collapses. When measuring the secondary (high voltage) side of the system, a specific secondary ignition pick-up lead must be used.

Figure 7-19 shows the waveform on the primary side of a traditional coil, of the type to be found on cars that use distributors. The measurement is made between the negative coil terminal and ground – this is the switched side of the coil. The tall vertical line shows the induced voltage (also known as primary peak volts) that occurs on the point of ignition – this reaches 400V on this screen

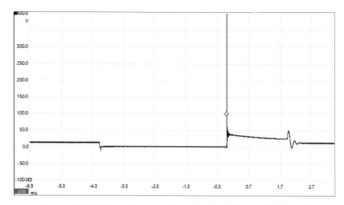

Figure 7-19: The primary side of an ignition coil. The tall vertical line shows the induced voltage that occurs on the point of ignition – this reaches 400V on this screen grab. (Courtesy Pico Technology)

81

WORKSHOP PRO CAR ELECTRICAL AND ELECTRONIC SYSTEMS

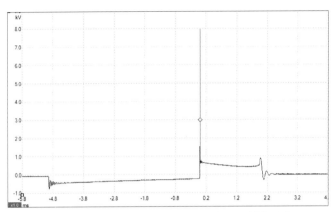

Figure 7-20: The secondary (high voltage side) of the ignition coil system. The waveform pattern is very similar to the primary side – but note that now the peak is at 8kV! (Courtesy Pico Technology)

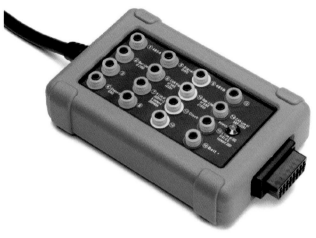

A CAN breakout box that plugs into the OBD port and allows easy access to the bus. (Courtesy Pico Technology)

grab. The sparkplug burn time extends from this spike to the start of coil oscillations – about 1.4ms here. There should be at least four oscillation peaks, counting both upper and lower. A loss of oscillations suggest that the coil should be changed. Figure 7-20 shows the secondary (high voltage side) of the system. The waveform pattern is very similar, but note the peak is at 8kV (ie 8000V)!

DATA BUSES

Data bus messages can be seen (but not interpreted as to actual content) with an oscilloscope. In some cases, these signals are available at the OBD port, while in other cases (and with other buses), the signals will need to be accessed at an ECU. Figure 7-21 shows Controller Area Network (CAN) bus signals, with CAN High shown in blue and CAN Low in red. As described in Chapter 5, you can see that the voltage signals are mirror images.

A Local Interconnect Network (LIN) bus scope screen grab is shown in Figure 7-22. LIN bus communication is a low-speed, single-wire serial data bus (a sub-bus of the faster, more complex CAN bus) used to control low-speed, non-safety-critical functions on the vehicle, especially windows, mirrors, locks, HVAC units, and electric seats. The lower level voltage (logic zero) should be less than 20% of battery voltage (typically 1V) and the upper level voltage (logic one) should be more than 80% of battery voltage. Note that the voltage levels may change slightly when the engine is started.

I use this colour Siglent SHS806 two-channel oscilloscope. This unit can also perform multimeter and logging functions. (Courtesy Siglent)

Figure 7-21: Controller Area Network (CAN) bus signals. Note that the time-base is in microseconds - μs. (Courtesy Pico Technology)

The K-Line is a very low-speed single-wire serial communication system used on many motor vehicles and commercial vehicles. It is commonly used for the diagnostic connections between the Electronic Control Modules (ECMs) on the vehicle and the diagnostic equipment (scan tools and data loggers). Figure 7-23 shows a K-line screen grab.

7. OSCILLOSCOPES

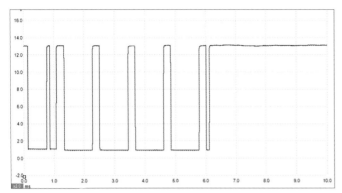

Figure 7-22: A Local Interconnect Network (LIN) bus scope screen grab. The time-base is in milliseconds – ms. (Courtesy Pico Technology)

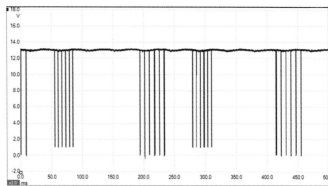

Figure 7-23: A K-line screen grab. The time-base is much longer than with the two previous bus signals. (Courtesy Pico Technology)

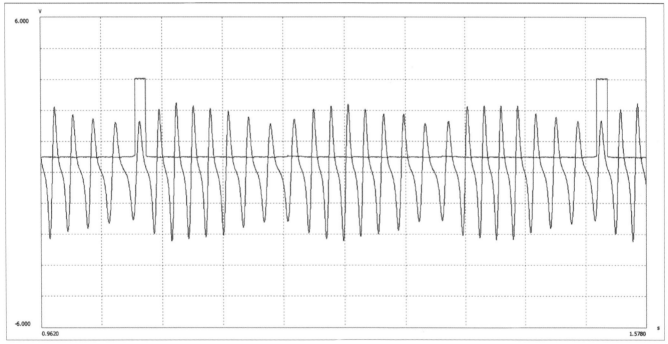

Figure 7-24: A screen grab from the inbuilt scope in a MoTeC programmable engine management system. It shows the outputs from an inductive crankshaft sensor and from a conversion module that simulates a Hall Effect output on the cam sensor. Programmable engine management systems often have a built-in scope feature.

WORKSHOP PRO CAR ELECTRICAL AND ELECTRONIC SYSTEMS

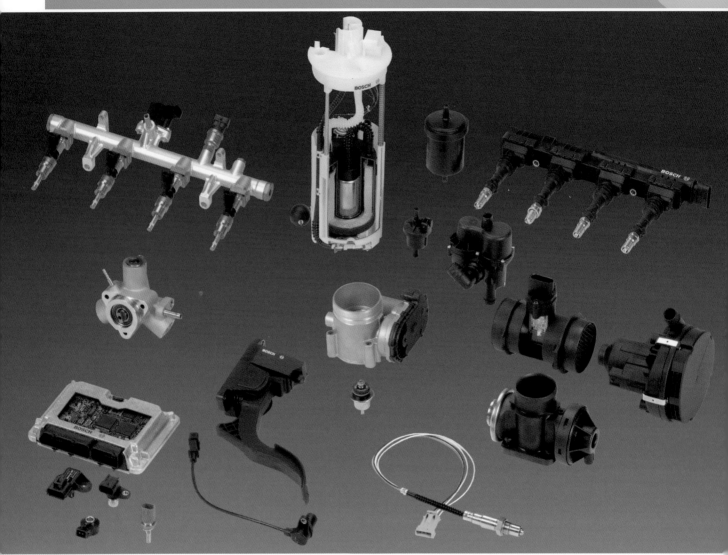
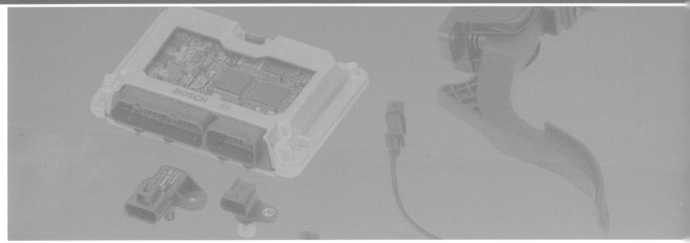

Chapter 8

Engine management

- Engine management system requirements
- Inputs and outputs
- Electronic Control Unit
- L-Jetronic
- Mono-Jetronic
- Motronic
- ME-Motronic
- Direct petrol (gasoline) injection
- Common rail diesel engine management
- Other engine management functions

WORKSHOP PRO CAR ELECTRICAL AND ELECTRONIC SYSTEMS

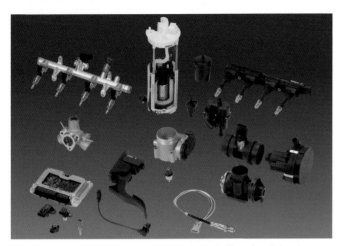

The components that make up a Bosch direction injection system. (Courtesy Bosch)

Electronic control of engine functions was first introduced in the mid-late 1970s. Initially, the ignition system was controlled by conventional means (eg a distributor working with points) and only the fuel function was made electronic. These cars were termed 'Electronic Fuel-injection' (EFI) cars – for example, early Bosch L-Jetronic systems worked in this way. The next step was for both fuel and ignition control to be become electronically controlled. Bosch Motronic is one of the most famous of the early systems that performed both fuel and ignition control. Subsequently, engine management systems adopted an increasing number of functions, for example controlling turbo boost, operating the electronic throttle and integrating with other control systems eg transmission control and electronic stability control.

However, the basics – fuel and spark – continue to be the most important aspects that engine management systems control, so I will start with those. The following discussion concentrates on port-injected, petrol (gasoline) systems – direct injection and diesel systems are discussedlater in this chapter.

ENGINE MANAGEMENT SYSTEM REQUIREMENTS

All engines require that the combustion air and fuel get mixed in the right proportions, and that each cylinder's sparkplug is fired at the right time.

The amount of air that is mixed with the fuel is described as the air/fuel ratio. A 'rich' mixture (eg 12:1) uses lots of fuel and is good for power, while a 'lean' mixture (eg 17:1) uses less fuel and so is more economical. With normal petrol (gasoline), an air/fuel ratio of 14.7:1 is termed 'stoichiometric' and is the chemically correct ratio for complete combustion. A stochiometric air/fuel ratio allows the catalytic converter to function best, and so give reduced emissions outputs. Many engine management systems now hold a fixed stochiometric air/fuel ratio, while older systems allowed the air/fuel ratio to vary within defined limits, depending on engine operating conditions.

Ignition timing refers to the point in the crankshaft's rotation at which the sparkplug fires. This is normally expressed in crankshaft degrees. If the spark is fired when the crankshaft is 15 degrees Before Top Dead Centre (BTDC), the spark timing is referred to as having '15 degrees of advance.' The ignition advance needs to vary because at high engine speeds, there is less time available for combustion to occur, so the spark needs to start the burn earlier. Changes in timing are also needed at different engine loads, different temperatures, and with different octane fuels.

THE INPUTS

In order that the engine's Electronic Control Unit (ECU) makes the right decisions about how much fuel to inject and when to fire the spark, it needs to know precisely the operating conditions of the engine. Is the car travelling at full throttle up a long hill at 3000rpm in very hot conditions? Or is the cold engine idling after it has been started for the first time that day? It is the input sensors that provide this type of information to the ECU. Let's now look at some of the most common input sensors.

Temperature sensors

The Coolant Temperature Sensor is used by the ECU to determine engine temperature. This sensor is normally mounted on the thermostat housing and comprises a device whose resistance changes with temperature. Usually as the sensor gets hotter, its resistance decreases. The sensor forms one resistance in a voltage divider circuit that is fed a regulated voltage by the ECU. The temperature of the intake air is also sensed, with this device located on the airbox or on an intake runner. Like the coolant temperature sensor, the Intake Air Temp Sensor is normally a temperature-dependent resistor. Some cars use other temperature sensors to measure fuel, cylinder head and oil temperatures.

An intake air temperature sensor. It is typically screwed through the wall of the intake manifold.

Airflow meters

Many cars use airflow meters to measure engine load. Airflow is relevant because the amount of power being

8. ENGINE MANAGEMENT

developed is proportional to the mass of air the engine is breathing per second. If the engine is drawing in a lot of air (for example, because the throttle is fully open at 6000rpm with the car travelling quickly up a steep hill), lots of fuel will need to be injected to keep the air/fuel ratio correct. The high load that the engine is undergoing will also influence the ignition timing that the ECU selects.

There are a number of different types of airflow meter available.

In a hot wire airflow meter the intake air flows past a very thin, heated platinum wire. This wire is formed into a triangular shape and suspended in the intake air path. The platinum wire forms one arm of a Wheatstone bridge electrical circuit and is maintained at a constant temperature by the electricity flowing through it. If the mass of air passing the wire increases, the wire is cooled and its resistance drops. The heating current is then increased by external circuitry to maintain the bridge balance. The value of the heating current (as measured by the voltage drop across a series resistor) is therefore related to the mass intake airflow.

Hot wire airflow meters have very quick response times and are internally temperature compensated. To make sure that the platinum sensing wire remains clean, it is heated red-hot for a short time after the engine is switched off, burning off any dirt or other contamination (note that in some airflow meters the hot wire is replaced with a hot-film resistor). Hot wire airflow meters can have either a 0-5V or variable frequency output signal. In the US, a management system using this approach is sometimes called 'MAF' – mass airflow.

One of the oldest automotive airflow meters is the vane airflow meter. This employs a pivoting flap which is placed across the inlet air path. Once the engine starts, low air pressure is experienced on the engine side of the vane, causing the flap to open a small distance. As engine load increases, the flap is deflected to a greater and greater extent. To prevent the flap overshooting its 'true' position (and also erratic flap movements being caused by intake pressure pulsing), another flap is connected at right angles to the sensing vane. This secondary flap works against a closed chamber of air, thus damping the motion of the primary vane.

Mechanically connected to the pivoting assembly in a vane airflow meter is a potentiometer, usually made up of a number of carbon resistor segments and a metal wiper. As the vane opens in response to the airflow, the wiper arm of the pot moves across the segments, changing the electrical resistance of the assembly. A regulated voltage is fed to the airflow meter, so that as the vane moves in response to airflow variations, the output voltage from the meter changes.

A spiral spring with an adjustable pre-load is used to

A hot-film airflow meter. All combustion air for the engine passes through this device.

relate the angle of the flap to the airflow, and to ensure that the flap closes when there is no airflow. An air bypass is constructed around the flap, with an adjustment screw positioned in this bypass, allowing control over idle mixtures. Because they measure only air volume (not mass), vane airflow meters require built-in temperature sensors. With both air volume and temperature inputs, the ECU can work out the mass of air being breathed.

A fuel pump switch is built into many vane airflow meters. This is designed to shut off the fuel pump if there is no air passing through the meter and so disables the pump when the vane is in its fully closed position. Because of the need to measure air volume, air temperature, and to operate the fuel pump, vane airflow meters can have up to seven connections within their plug. Vane airflow meters typically have 0-5V signal outputs, but some use a 0-12V output range.

Karman Vortex airflow meters generate vortices whose frequencies are measured by an ultrasonic transducer and receiver. They use a flow-straightening grid plate at the inlet to the meter and can be quite restrictive. This type of meter has a variable frequency output. Karman Vortex airflow meters are in some respects an oddity in airflow meter design – they were used on a relatively small number of makes and models.

MAP sensors

Many cars don't use any form of airflow meter, instead relying on the input of three sensors to work out how much air is being breathed by the engine. One is the inlet air temperature sensor, the second is an engine speed sensor

A MAP sensor. This one is designed to bolt to the intake manifold.

(I'll get to that in a moment), and the third is a manifold vacuum sensor – this sensor is called a MAP (Manifold Absolute Pressure) sensor. A MAP sensor continuously measures the pressure in the intake manifold.

The pressure in the intake manifold is dependent on rpm, throttle opening and load. In a naturally aspirated car, the measured pressure will be below atmospheric in all conditions except full throttle (where it will be much the same as atmospheric pressure). In a car with forced aspiration (a turbocharged or supercharged car), the pressure will be above atmospheric when the engine is on boost, and below atmospheric when it is not. In a naturally aspirated car, when the MAP sensor registers a low manifold vacuum and the rpm sensor indicates that the engine is at high rpm, a high load situation is signalled to the ECU. High manifold vacuum at high *or* low revs means that the throttle is closed – indicative of a low load situation. In the US, the MAP sensor approach is sometimes called 'speed density.'

A MAP sensor is fed manifold pressure via a port situated after the throttle butterfly. MAP sensors are usually mounted near (or on) the engine, but in some cars the MAP sensor is located within the ECU.

Crankshaft and camshaft position sensors

The ECU needs to know how fast the engine crankshaft is rotating, and where it is in its rotation. It uses one or more position sensors to collect this information. Most cars use both camshaft and crankshaft position sensors, allowing the ECU to sequentially fire the spark and injectors. Older cars that did not use sequential firing tended to use just a crankshaft position sensor.

There are many different types of position sensors.

Optical position sensors use a circular plate with slots cut into it. The plate is attached to the end of the camshaft and is spun past a nLED. A sensor on the other side of the disc registers when it sees the light shining through one of the slots, with the ECU then counting the light pulses. Some optical sensors use 360 slots in the disc, allowing very fine resolution of engine speed. Optical sensors were widely used in the 1980s but are now seldom seen.

A Hall Effect position sensor uses a set of ferrous metal blades that pass between a permanent magnet and a sensing device. Every time the metal vane comes between a magnet and the Hall sensor, the Hall sensor switches off. This gives a signal whose frequency is proportional to engine speed.

An inductive position sensor reads from a toothed cog. It consists of a magnet and a coil of wire, and as a tooth of the cog passes, an output voltage pulse is produced in the coil. Both inductive and Hall Effect sensors have been widely used over a long period.

A crank angle sensor.

Oxygen sensor

The oxygen sensor (sometimes called the Lambda Probe) is located in the exhaust. It measures how much oxygen there is in the exhaust compared with the atmosphere and so indicates to the ECU the air/fuel ratio.

Two different types of oxygen sensor have been widely used. The first is called a 'narrow band' sensor. This sensor generates its own voltage output, just like a battery. When the air/fuel ratio is lean, the sensor emits a very low voltage output, eg 0.2V. When the mixture is rich, the voltage output is higher, eg 0.8V. The ECU uses the output of this sensor to keep mixtures around 14.7:1 in cruise and idle conditions. To facilitate this, the voltage output of the sensor switches quickly from high to low (or low to high) as the mixtures move through the 14.7:1 stoichiometric ratio. Older narrow band oxygen sensors use just a single wire (and engine earth). These sensors are unheated, while later three-wire sensors have a 12V heating element within them.

More recent cars use wideband sensors. These are able to measure air/fuel ratios over a wide range. This allows the ECU to remain in closed loop (more on closed

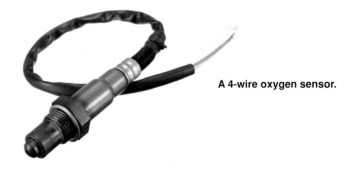

A 4-wire oxygen sensor.

88

8. ENGINE MANAGEMENT

loop in a minute) the whole time, using the input of the wideband sensor to measure air/fuel ratios in all conditions except cold start. Unlike a narrow-band sensor, wideband sensors have a voltage output that is more linear with respect to air/fuel ratio. Wideband sensors use four or even five connections.

Oxygen sensors can be contaminated by the use of inappropriate silicone sealants and leaded petrol. They can also become carboned-up over time, slowing their response.

Many cars use multiple oxygen sensors, positioned before and after the catalytic converter(s). Cars equipped with sensors arranged in this manner have ECUs that can assess the effectiveness of the cat converter operation. Twin turbo and 'V' engines usually have at least two oxygen sensors, and may have as many as four.

Knock sensor

The knock sensor is like a microphone listening for the sounds of knocking (detonation). It is screwed into, or bolted to, the block, and works in conjunction with filtering and processing circuitry in the ECU to sense when knocking is occurring. Overly advanced ignition timing for the prevailing engine conditions will cause knocking. Many engines with engine management run ignition timing very close to knocking, so that the information provided by this sensor is vital if engine damage isn't to occur on a bad batch of fuel or on a very hot day. Some cars (especially those with V engines) run four knock sensors. Some systems can isolate individual cylinder knocking and retard the timing on only that cylinder.

A knock sensor. It is attached to the engine via a large bolt that passes through the middle opening.

Throttle Position Sensor

The Throttle Position Sensor (TPS) indicates how far open the throttle blade is. Throttle position sensors on older cars comprise a pot whose wiper is moved by the opening of the throttle. The pot is fed a regulated voltage (usually 5V) and the output signal varies from typically about 0.8V to 4.5V, with a higher voltage indicating a greater throttle opening. These throttle position sensors are usually mounted on the throttle body and have a three-wire plug.

In cars with electronic throttle control, two position

A throttle position sensor. Internally, it is a potentiometer.

sensors are used on the throttle body. The ECU looks at the outputs of both sensors to ensure that they are consistent with one another, giving greater certainty that the signals are accurate; the two signals usually work in opposite directions. The driver's requested throttle position is assessed not by the throttle position sensors, but by accelerator pedal sensors. These can be mounted on the accelerator pedal (this is most common) or still be on the throttle body – the latter used if a throttle cable is retained.

Very old cars with EFI may have a throttle sensor that comprises just two switches – full throttle and idle.

Other input sensors

There are other system inputs that can be used. These can include, but are not limited, to the following:
- Vehicle speed sensor
- Atmospheric pressure sensor
- Fuel pressure
- Water temperature, vacuum, brake and clutch switches
- EGR position sensor
- A neutral position switch

In addition, the ECU is modern cars is likely to gain a lot of information from other sensors and systems via data bus connections (eg the CAN bus).

A vehicle speed sensor.

The outputs

The most important components that the ECU controls are the injectors. The amount of fuel that flows through the injectors is determined by the length of time that the ECU switches them on. This opening time – called the pulse width – is measured in milliseconds. When an injector is open, fuel squirts from it in a fine spray.

While it varies from car to car, the injectors in many

cars squirt once for each two rotations of the crankshaft. In older cars, injectors are often fired in banks of three of four at a time, while in more modern cars, the injectors are triggered sequentially to match the cylinder firing order.

The percentage of time that the injectors are open for is called the duty cycle. An injector open for half the time has a 50 per cent duty cycle, while if it is open for three-quarters of the available time, it has a 75 per cent duty cycle. At maximum power in a standard engine, the injectors might have an 85 per cent duty cycle. That means the injectors are flowing at 85 per cent of their full capacity.

Injection can be of two quite different types. In port injection systems, the injectors flow fuel in a fine spray onto the back of the inlet valves; that is, they squirt into the inlet port. In direct injection engines, the fuel is squirted into the combustion chamber itself. Some engines mix both types of injection; that is, they feature both port and direct injection approaches.

In addition to controlling the injectors, the ECU in engine-managed cars also controls the ignition timing. However, in most cars, the ECU doesn't directly control the ignition coil. Instead, an ignition module is used to switch the power to the coil(s) on and off; the ECU tells the ignition module exactly when it needs to switch to provide the correct ignition timing.

Older engine management systems are equipped with a single coil and a distributor. The distributor uses a rotor spinning around inside the cap to send high voltage electricity to the sparkplug at the right moment. In this type of system, the crankshaft position sensor is usually located inside the distributor body. However, most cars of the last 20 years use a multi-coil arrangement which can involve either a single coil for each sparkplug, or double-ended coils. These direct fire ignition systems do not use a distributor because the coils are fired individually. With double-ended coils, two of the engine's spark plugs are fired at the same time – one on a cylinder that is on the compression stroke (which does something), and one on the cylinder on an exhaust stroke (which doesn't do much!). On these engines, the crankshaft position sensor

An ignition coil. This is a coil that fits directly on the sparkplug. It has switching electronics built into the top of the coil assembly.

is usually mounted on the crank itself. Many cars have the coils mounted directly on the sparkplug, so removing the need for high tension leads. These coils have the switching electronics built into the coil assemblies.

Idle speed control is carried out by changing the amount of air that can bypass the nearly closed throttle butterfly. Some cars used a pulsed valve (a little like an injector in the way it switches on and off) to regulate the amount of air that can bypass the throttle body. If the idle speed needs to be lifted, the duty cycle of the valve is increased and more air passes through. Stepper motors are also used in some systems. In cars with electronic throttle control, the idle speed is set by the system directly operating the throttle.

Older cars often used an additional valve to control idle speed when the engine is not yet up to operating temperature. Typically, the valve (called an auxiliary air control valve) has engine coolant passing through its body and is also equipped with an electric heating element. As it warms, an open passage through the valve gradually closes, reducing the amount of air bypassing the throttle. This allows more air to be breathed by the engine when it is cold, keeping the idle speed high. Other engines use 'idle-up' solenoids that allow extra intake air to bypass the throttle when loads such as the air conditioner are switched on.

THE ELECTRONIC CONTROL UNIT

Rather than trying to understand the functions of each component in an ECU, it is better to think about the control logic that an ECU uses to run an engine. Here I will look at some of the general strategies adopted by the ECUs in most cars.

Closed and open loop

Closed and open loop refers to the strategy that the ECU is using to maintain the air/fuel ratio at the desired level.

A port fuel injector. When power is applied, the injector opens atnd fuel flows.

8. ENGINE MANAGEMENT

In closed loop control, the ECU is using the input of the oxygen sensor to measure the air/fuel ratio. If the air/fuel ratio is incorrect, the ECU makes an appropriate adjustment to fuelling. In open loop, the ECU sets the pulse width of the injectors according to a pre-programed map – and doesn't check the result.

Closed loop on older cars meant that the air/fuel ratio was stoichiometric, that is, it was hovering around 14.7:1. The narrow band sensors that were used in these cars were really only able to detect this condition. However, cars fitted with wideband sensors can detect the air fuel ratio in all operating conditions, and so the input of the oxygen sensors can be maintained constantly. Thus a car with a wideband oxygen sensor can be in closed loop the whole time, and have appropriately varying air/fuel ratios for load or other conditions.

Fuel trims
In addition to closed loop running, the oxygen sensor is also used as part of the ECU's self-learning system. Imagine for a moment that the fuel filter in your car is a little blocked, causing the mixtures to be always a little lean. The oxygen sensor measures this and in response, the ECU enriches the air/fuel ratio. But it's a pretty inefficient system if every day the ECU has to respond in the same way! Instead, what happens is this. The ECU knows that the mixtures are always a bit lean, and so it *permanently* enriches the mixtures. It has 'learned' that richer mixtures are needed, and so always runs this correction. If you change the filter, the mixtures will be a little rich until the ECU gradually re-learns the new requirements. This self-learning process occurs in most ECUs and is totally dependent on the health of the oxy sensor(s). This change of fuelling strategy over time through self-learning is often called the 'fuel trim.' ECUs can display, via OBD, their short- and long-term fuel trims.

Lean Cruise
We already know that, ignoring emissions for the moment, rich mixtures are needed for power and leaner mixtures for cruise. But what about a long, gentle drive? Even leaner mixtures can then be used, improving economy further. And that's just what a 'lean cruise' ECU does. It takes note of how long the car has been maintaining a steady speed for, how much throttle is being used, and whether the engine coolant is up to operating temperature. If all of these factors are OK, the ECU will start to lean out the mixtures. Second by second the air/fuel ratio will gradually become leaner, until the engine is running at an air/fuel ratio as lean as 18 or 19:1! If you put your foot down, the ECU instantly forgets all about lean cruise – until the right conditions are again met. Not all cars run a lean cruise mode, as emission of oxides of nitrogen rise rapidly in

A Toyota engine management ECU.

this mode, but for those that do, fuel economy can be considerably improved.

Engine speed limiting and over-run fuel cut-off
All engine management systems use a rev limiter. Some limiters cut off fuel completely at the prescribed engine speed, withholding it until you're 500 rpm below the limit. Hitting this 'bed of nails' limiter makes you think that you've just broken the crankshaft! Other rev limiters cut off the spark or injectors of individual cylinders one after the other, so that you can barely feel that you have reached the maximum allowable rpm. These soft limiters mean that the car can be used right to the rev limit without a worry.

When zero throttle is used at higher speeds, the ECU switches off the injectors. For example, this situation occurs when you are approaching a red traffic light on a main road and have lifted the throttle completely. The injectors resume flow only when engine revs drop to around 500 rpm above idle. This injector shut-off benefits both fuel economy and emissions and is one of the ECU outcomes reliant on the input of the road speed sensor.

Limp home
It sounds strange to hear it now, but when engine-managed cars were first released, it was widely suggested that they would be very unreliable – obviously, this has proved not to be the case. One reason for this is that the ECU has internal back-up values and strategies to adopt if a sensor fails. For example, if the coolant temperature sensor becomes defective (or the wire to the sensor is damaged), the ECU is programmed to ignore the incorrect input. Instead of measuring coolant temp, the ECU might

WORKSHOP PRO: CAR ELECTRICAL AND ELECTRONIC SYSTEMS

rely only on the intake air temp sensor. Alternatively, the ECU might replace the coolant temperature sensor's input with a pre-programmed value. When an ECU does this, it is said to be in 'limp home' mode.

Years ago, I did an experiment where emissions of a car were being measured on a dyno. As part of the test, we disconnected various engine management sensors and then checked on how this impacted emissions. The car – a Honda with standard emissions well below the legal requirement – was able to cope with an extraordinary number of sensors being disconnected before it finally produced illegal emissions.

Self-diagnosis and On-board Diagnostics
All cars of the last 25 years have some type of self-diagnosis capability built into their engine management systems. These are covered in more detail in Chapter 10.

EXAMPLE ENGINE MANAGEMENT SYSTEMS
I'd like now to look at some engine management systems in more detail. It is impossible to cover all systems that have been used but the following systems (or their derivatives) have been used in an enormous number of cars, and will give you a good indication of how a range of systems work. All the described systems are from Bosch, a company that has an exemplary record in publishing technical information on their equipment.

BOSCH L-JETRONIC – ELECTRONIC FUEL-INJECTION
Bosch L-Jetronic fuel-injection was one of the earliest electronic fuel-injection systems widely fitted to production cars. In addition to being sold under the Bosch name, very similar licensed systems were used on millions of other cars. Figure 8-1 shows the layout of a later L-Jetronic system. L-Jetronic was fitted to cars from 1974-1989.

L-Jetronic systems had analog ECUs. The original Bosch description of the ECU proudly states that it has no less than 250 components, including 30 transistors and "nearly 40 diodes!" The ECU connection plug had only 25 pins. My second car – a 1977 BMW 3.0si – used L-Jetronic fuel-injection, and even though I owned it when it was a decade old, the system was still seen as being advanced for the time ... and, compared with carburettors, it was!

Injectors in the earliest L-Jetronic systems were fired in groups – a pair at a time in the case of four-cylinder engines and three at a time with six-cylinder engines. (There were no V8 L-Jetronic cars then or, as far as I know, later either.) Fuel timing was provided by an input from the distributor. In addition to the normal points used for ignition timing, a second set of contacts fed speed and timing information to the ECU. Some L-Jetronic systems used a bulky MAP sensor, but most cars used a vane airflow meter. Coolant and (in some systems) intake air temperatures were measured by sensors.

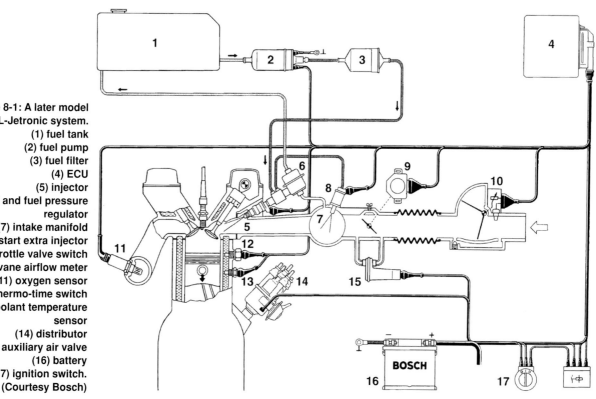

Figure 8-1: A later model L-Jetronic system.
(1) fuel tank
(2) fuel pump
(3) fuel filter
(4) ECU
(5) injector
(6) fuel rail and fuel pressure regulator
(7) intake manifold
(8) cold start extra injector
(9) throttle valve switch
(10) vane airflow meter
(11) oxygen sensor
(12) thermo-time switch
(13) coolant temperature sensor
(14) distributor
(15) auxiliary air valve
(16) battery
(17) ignition switch.
(Courtesy Bosch)

8. ENGINE MANAGEMENT

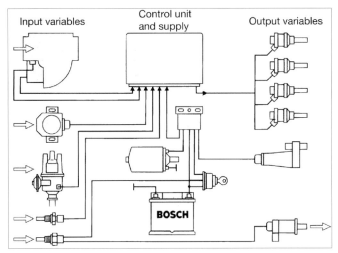

Figure 8-2: Schematic layout of an L-Jetronic system. On the left are ECU inputs from the vane airflow meter, throttle position switch, distributor contacts, and coolant and intake air temperatures. At right are the ECU outputs – the injectors, auxiliary air valve and cold-start injector. (Courtesy Bosch)

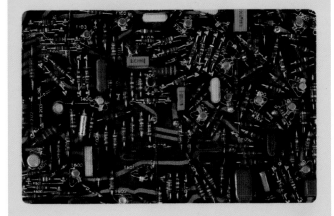

The cover of this Bosch technical instruction book features a photo taken inside an early L-Jetronic ECU. (Courtesy Bosch)

Without the ability to internally apply digital look-up table corrections for coolant temperature and full-load enrichment, these areas of control were performed quite differently to later systems. An extra enrichment injector sprayed-in fuel on a cold start. This was a continuously flowing injector that was controlled by a time or thermo switch. An auxiliary air valve provided the increased air needed to go with this fuel, with this valve closing an orifice as the valve was heated by the engine coolant. Full-load enrichment was triggered by a throttle or manifold pressure switch. Even this very early system had over-run fuel injector cut off and acceleration enrichment provisions. Later L-Jetronic systems added closed-loop control via a narrow-band oxygen sensor.

These systems have no form of self-diagnosis. Therefore, fault-finding has more in common with the approaches covered in Chapter 4 (Fault-Finding Basic Car Electrical Systems) than Chapter 10 (Fault-Finding Advanced Car Systems). For that reason, I'll cover some simple L-Jetronic fault-finding here.

If the car is experiencing starting problems, first make a thorough check of the ignition system and all hoses. The next step is to check fuel pressure, and then if that is OK, the action of the cold start injector. The thermo time switch (which operates the start injector), and then the auxiliary air control valve (which flows the increased air needed when the engine is cold), should then be checked. The vane airflow meter is the next step: can the vane be smoothly opened by hand without catching or sticking?

If the engine easily stalls, check the idle contact in the throttle position switch. The cold start injector should not be leaking, and the auxiliary air control valve should be closed when the engine is at operating temperature. Are the injectors working correctly, not dribbling or leaking?

A missing engine is likely to be either a problem in the ignition system or a faulty plug or connector in the L-Jetronic system. One unusual check in this situation is of the alternator: high or low voltages may cause misses. A miss under load may indicate a lower than required volume of fuel is being delivered by the pump. This can be checked by detaching the hose at the start valve and placing this hose in a container. Opening the airflow meter flap by hand will start the pump (the ignition switch needs to be on as well); after 1 minute expect 1-1.5 litres of fuel. A partially blocked injector can also cause a high-load miss. To test this, remove one injector at a time from the intake manifold and allow it to spray into a graduated container at idle. All injectors should flow an equal amount. Finally, of course, if the engine is missing, the ECU could be the issue: this can only be checked by replacing it with a known good unit.

Note that it is easy to change the air/fuel ratios of an L-Jetronic car that doesn't use an oxygen sensor (ie earlier L-Jetronic vehicles). If you have access to an air/fuel ratio meter, and the car is running lean (or rich) right through the

93

WORKSHOP PRO CAR ELECTRICAL AND ELECTRONIC SYSTEMS

A mid-Eighties BMW 735, which uses L-Jetronic electronic fuel-injection. The vane airflow meter can be seen on the far side of the engine bay, to the right of the air cleaner and just before the throttle body.

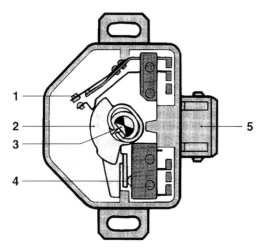

Figure 8-3: Throttle position switch from an L-Jetronic system. (1) full-load contact, (2) switching cam, (3) throttle valve shaft, (4) idle contact, (5) connector. Note that a potentiometer is not used. (Courtesy Bosch)

load range, the tension of the return spring in the airflow meter can be adjusted. This is achieved by removing the plastic cover, and then moving the spring adjustment mechanism – reducing preload if the car has been running lean, and increasing preload if the car has been running rich. Be very careful to mark your initial position so that you do not lose your place, and make only very small changes before testing the results.

BOSCH MONO-JETRONIC – A SINGLE-POINT INJECTION SYSTEM

As described earlier, the majority of EFI systems use one injector for each of the engine's cylinders. But there's another approach that's been taken to electronic injection – single point injection. These systems, which normally use only one or two injectors, allow the injector count to be more than halved, and an airflow meter or MAP sensor is not needed. The result is that the cost of the induction system can be made very low. But if the car is still to perform adequately, complex engineering solutions are needed. This section looks at a simple, cheap EFI system using just one injector and only four major input sensors. Again, it – and its derivatives – were widely fitted to cars, this time from about 1988 – 1995.

System layout

On paper, Mono-Jetronic appears similar to any of the more common EFI systems. Fuel is pressurised by an electric pump, fed through a fuel filter and then fixed at given headroom above the manifold air pressure by a fuel pressure regulator, before being fed to an electronically-controlled injector. Induction air passes through a filter, is monitored for temperature by an intake air temp sensor, and then passes through the throttle body into the engine. Figure 8-4 shows the overall system.

However, as Figure 8-5 shows, the physical layout of the system is quite unusual. The fuel injector, air temp sensor, fuel pressure regulator, throttle valve, idle speed control actuator and throttle position sensor are all integrated into one unit. Combining the various components into one package in this way reduces manufacturing and installation costs. The assembly is positioned in a similar location to that used by a carburettor in an old car – on top of a multi-branch intake manifold.

Collecting engine data

The two major inputs determining the injector pulse width are engine speed and throttle position. Engine speed is easily derived by the monitoring of the ignition signal, but the accurate sensing of throttle position is more difficult. When load sensing occurs through the monitoring of throttle angle, the relationship between throttle valve opening and the flow area within the throttle body must be

8. ENGINE MANAGEMENT

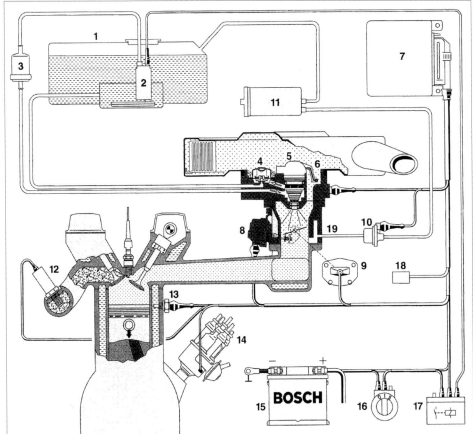

Figure 8-4: The Mono-Jetronic system.
(1) fuel tank
(2) fuel pump
(3) fuel filter
(4) fuel pressure regulator
(5) fuel injector
(6) air temperature sensor
(7) ECU
(8) throttle valve actuator
(9) throttle position sensor
(10) canister purge valve
(11) carbon canister
(12) oxygen sensor
(13) coolant temperature sensor
(14) distributor
(15) battery
(16) ignition switch
(17) relay
(18) diagnostics connector
(19) central injection unit.
(Courtesy Bosch)

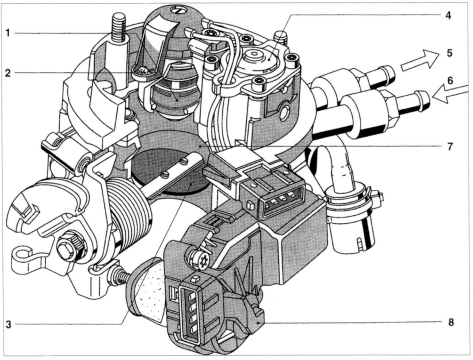

Figure 8-5: Mono-Jetronic central injection assembly.
(1) fuel injector
(2) intake air temp sensor
(3) throttle butterfly
(4) fuel pressure regulator
(5) fuel return
(6) fuel inlet
(7) throttle position sensor (hidden)
(8) idle air bypass motor.
(Courtesy Bosch)

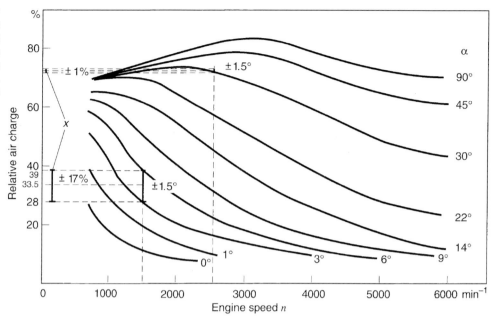

Figure 8-6: This graph shows air charge at different throttle angles and rpm. Note how the amount of air that the engine breathes varies widely with variations at very small throttle angles. (Courtesy Bosch)

maintained to within very close tolerances on all production units. This is because small throttle movements can make a huge change to engine load.

The first step in developing the system is to subject the engine to accurate dynamometer testing. This is so that the air charge for one intake cycle at various engine speeds and throttle openings can be measured. Figure 8-6 shows an example of these 'air charge' amounts.

Several interesting aspects can be noted about the diagram. Firstly, the amount of air breathed per intake stroke is at its maximum at peak torque, as is shown by the air charge line indicative of full throttle (the butterfly open by 90 degrees). As can be seen, the greatest ingestion per intake stroke occurs on this engine at about 3000rpm. However, of more importance when attempting to measure the correct amount of fuel needing to be added, are the differences in air charge amount which occur at small throttle openings. At idle and low-load, a change of ±1.5° throttle opening causes an air charge difference of ±17%. On the other hand, the same amount of throttle movement at high loads can cause a change of only ±1%. Obviously, then, small throttle openings must be measured with extreme accuracy.

In the Mono-Jetronic system a special throttle position sensor is used. The throttle position sensor uses dual potentiometers – one to cover throttle openings from 0-24°, and the other from 18-90°. Figure 8-7 shows this sensor. Track 1 covers the angular range from 0-24°, while Track 2 covers the range from 18-90°. The angle signals from each track are converted by dedicated analog/digital converter circuits. The ECU also evaluates the voltage ratios, using this data to compensate for wear and temperature fluctuations at the pot.

Because the engine load cannot be assessed in this way as accurately as by MAP sensing or airflow metering, the system requires the feedback of an exhaust gas oxygen (EGO) sensor if it is to comply with emissions legislation. The EGO sensor is of the narrow-band type, where the sensor output is a small voltage which changes rapidly in level either side of the stoichiometric air/fuel ratio. Other sensor inputs include coolant and intake air temp, and control signals from the air-conditioning and/or automatic transmission. The latter two inputs are used as part of the idle speed control strategy.

A cutaway view of the throttle body of the Mono-Jetronic system. (Courtesy Bosch)

8. ENGINE MANAGEMENT

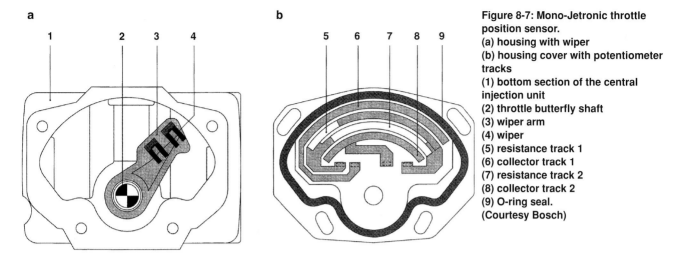

Figure 8-7: Mono-Jetronic throttle position sensor.
(a) housing with wiper
(b) housing cover with potentiometer tracks
(1) bottom section of the central injection unit
(2) throttle butterfly shaft
(3) wiper arm
(4) wiper
(5) resistance track 1
(6) collector track 1
(7) resistance track 2
(8) collector track 2
(9) O-ring seal.
(Courtesy Bosch)

Processing input data

Figure 8-8 is a schematic diagram of the system's ECU. The inputs from the TPS, EGO, engine temperature and intake air temperature sensors are converted to data words by the analog to digital converter and transmitted to the microprocessor by the data-bus. The microprocessor is connected through the data and address bus with the EPROM and RAM. The read memory contains the program code and data for the definition of operating parameters. In particular, the RAM stores the adaptation values developed during self-learning, which occurs on the basis of the EGO sensor input. This memory module remains permanently connected to the vehicle's battery to maintain the adaptation data whenever the ignition is switched off.

A number of different output stages are used to generate the control signals for the fuel injector, the idle speed control actuator, the carbon canister purge valve (which allows the burning of stored petrol tank vapour), and the fuel pump relay. The fault lamp warns the driver of sensor or actuator problems, and also acts as a diagnostics interface (ie when triggered appropriately, it flashes fault codes).

Mixture formation

The starting point for the calculation of the appropriate fuel injector pulse width is a stored three-dimensional map derived from dyno test data. This 'Lambda Map' (see Figure 8-9, overleaf) contains the optimum pulse widths

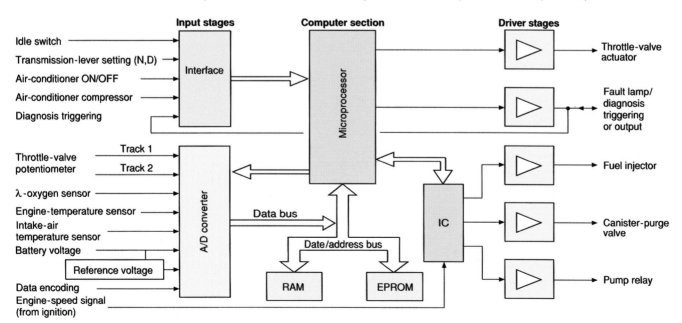

Figure 8-8: Mono-Jetronic inputs and outputs. (Courtesy Bosch)

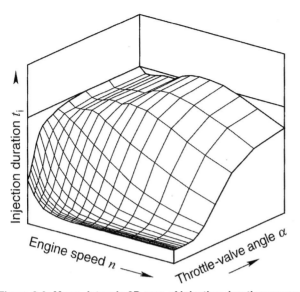

Figure 8-9: Mono-Jetronic 3D map of injection duration versus throttle position and engine speed. (Courtesy Bosch)

to deliver a stoichiometric air/fuel ratio under all operating conditions. The map consists of 225 control co-ordinates, made up of 15 reference co-ordinates for throttle position, and 15 for engine rpm. Because of the extremely non-linear shape of the air-charge curves, the data points are situated very closely together at the low-load end of the map. The ECU interpolates between the discrete points within the map.

If the ECU registers deviations away from stoichiometric air/fuel ratios, and as a result is forced to correct the basic injection duration for an extended time, it generates mixture correction values and stores them as part of the adaptation process. In this way engine-to-engine variations and engine wear are compensated for.

Compensations
Because the Mono-Jetronic system uses just a single injector location, manifold wall wetting through condensation is a much more major problem than in multipoint systems. As in all digital EFI systems, injector pulse width is increased when the engine is cold. However, because condensation of the fuel also depends on the air velocity, the starting injector duration is reduced as engine speed increases. To counteract the possibility of flooding, the longer the engine cranks, the less fuel that is injected, with it reduced by 80 per cent after six seconds of cranking. Once the engine has started, the injector opening duration is based on the values stored within the Lambda map, suitably modified on both a time and temperature basis by the engine coolant temperature input.

While all EFI systems use the equivalent of a carburettor accelerator pump during rapid throttle movements, the single injector location of the Mono-Jetronic system makes this a critical area. During sudden changes in throttle position three factors need to be taken into consideration:
- Fuel vapour in the central injector unit and intake manifold is transported very quickly – at the same speed as the intake air.
- Fuel droplets are generally transported at the same speed as the intake air, but are occasionally flung against the intake manifold walls, where they form a film which then evaporates.
- Liquid fuel is transmitted as a fuel film on the intake manifold walls, reaching the combustion chambers after a time lag.

At idle and low loads, the air pressure within the manifold is low (ie there is a high vacuum), and the fuel is almost entirely vapour with no wall wetting. When the throttle valve is opened, the intake manifold pressure rises, and so does the proportion of fuel on the manifold walls. This means that when the throttle is opened, some form of compensation is necessary to prevent the mixture becoming lean due to the increase in the amount of fuel deposited on the walls. When the throttle is closed, the wall film reduces, and without some form of leaning-compensation the mixture would become rich!

Rather than basing the transitional compensation on throttle position alone, the system uses the *speed* with which the throttle is opened or closed as the determining factor. Maximum correction occurs when the throttle is opened at more than 260 degrees per second. Also incorporated in these dynamic mixture corrections are the input of the engine and intake air temperature sensors.

Mixture adaptation
The mixture adaptation system uses the EGO sensor input. The system must compensate for three variables:
- air-density changes when driving at high altitudes
- vacuum leaks after the throttle butterfly
- individual differences in injector response times

The frequency of the updates varies between 100 milliseconds and 1 second, depending on the engine load and speed.

BOSCH MOTRONIC – IGNITION AND FUEL CONTROL
When most people think of standard engine management systems, they are likely to be thinking of the Bosch Motronic system or one of its of close derivatives. Engine management – the control of *both* fuel and spark – really came about with this system. It is an approach that is still widely used today. Motronic uses digital ECUs that contains 'maps' – three-dimensional look-up tables that allow the controlled output (for example injector pulse width or ignition timing) to be set on the basis of load and rpm.

8. ENGINE MANAGEMENT

With a look-up table, any point on the map can have any output that's desired.

When Motronic was released, Bosch stated that the advantages of the system were:

- Fuel savings, especially in comparison with carburettors and points ignition, but even compared to engines with gasoline injection and conventional transistor ignition.
- Fuel savings because of precisely metered mixture enrichment during warm-up and appropriate adjustment of ignition timing.
- Fuel savings through precisely metered, rpm-dependent mixture enrichment for full-load operation.
- Overrun fuel cut-off that further reduces fuel consumption.
- Reduction of fuel consumption while meeting emissions regulations through the tailoring of ignition advance to all operating conditions.
- Optimum ignition advance and fuel metering during starting, which gave dependable starting and cold-starting conditions.

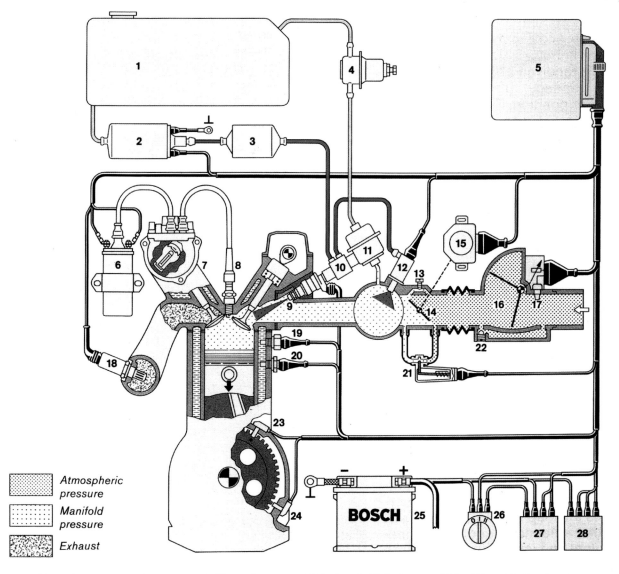

Figure 8-10: An overview of an early Motronic system. (1) fuel tank, (2) fuel pump, (3) fuel filter, (4) vibration damper [often mis-identified as the fuel pressure regulator], (5) ECU, (6) coil, (7) distributor, (8) sparkplug, (9) injector, (10) injector rail, (12) cold start valve, (13) idle speed adjusting screw, (14) throttle valve, (15) throttle valve switch [not yet a pot on early Motronic], (16) vane airflow meter, (17) intake air temperature sensor, (18) oxygen sensor, (19) thermo time switch, (20) coolant temperature sensor, (21) auxiliary air valve, (22) idle mixture adjustment screw, (23) crank timing sensor, (24) crank speed sensor, (25) battery, (26) ignition switch, (27) main relay, (28) fuel pump relay. (Courtesy Bosch)

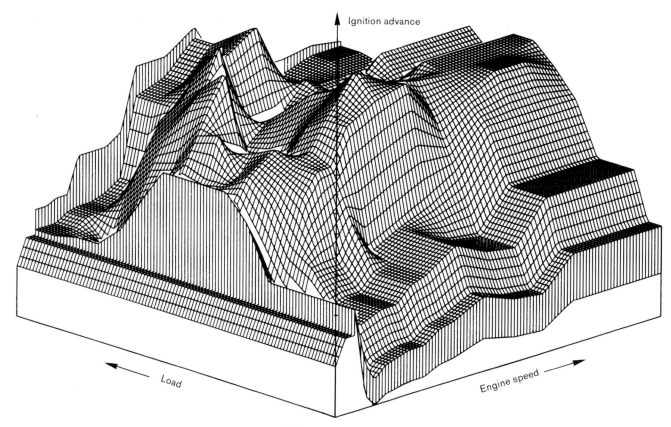

Figure 8-11: The classic three-dimensional map of the digital engine management system. This represented a revolution at the time: any value of ignition timing or fuel injector pulse width could be programmed for any combination of engine speed and load. This map shows ignition advance versus load and speed. (Courtesy Bosch)

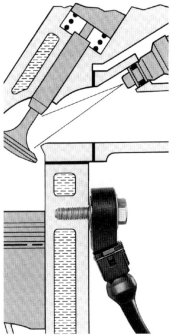

Figure 8-12: Motronic added ignition timing to fuel injection control. This led to the development of the knock sensor, shown here bolted to the block. This diagram also shows very clearly the injector location in port fuel-injection systems (top). (Courtesy Bosch)

- Stable idling.
- Good torque characteristics at low engine speeds.
- Full-load ignition advance set for best torque, and set for minimum fuel consumption while meeting emissions limits in part-load ranges.
- Low emissions through excellent tailoring of fuel quantity and ignition advance to load conditions.

And this was just part of it – the list went on! It wasn't just marketing, either (perhaps with the exception of over-run fuel cut-off, which was available even on L-Jetronic). The ability to precisely control ignition advance in a way that's simply impossible with a conventional vacuum and centrifugal advance system, and the move to full digital control, gave results that were a huge leap forward over previous technologies. As a specific example, it allowed the use of turbocharging in a way that was previously difficult or even impossible. (If you don't remember, look at the primitive ignition and fuel control systems being used on just the handful of turbo cars that were produced before engine management became available.) Motronic became available in the mid-1980s and was used for about 20 years.

8. ENGINE MANAGEMENT

In many respects, the first Motronic systems looked very much like the last L-Jetronic systems. They used a vane airflow meter, a fuel supply system and injectors that, if not identical to late L-Jetronic systems, were certainly very similar. Many even retained distributor-type ignition systems. Figure 8-10 (page 99) shows the layout of an early Motronic system. Note the use of the L-Jetronic cold-start and auxiliary air control valves, and vane airflow meter. As time passed, these cold start valves were replaced with logic that provided cold-start enrichment through the standard injectors, the auxiliary air control valve was replaced with a PWM-controlled idle speed control valve, and the vane airflow meter replaced with a hot-wire (and later, hot film) airflow meter. Direct ignition, with a coil typically mounted on each plug, replaced the distributor.

However, the fundamental control logic of the Motronic system – using three-dimensional look-up tables, able to expressed as a 3D graph (see Figure 8-11) – stayed in place for decades. The approach is still used today in programmable engine management systems.

BOSCH ME-MOTRONIC – ELECTRONIC THROTTLE CONTROL AND TORQUE MODELLING

The Bosch ME-Motronic system takes a quite different approach to the engine management systems described so far. While at first it appears to have only the usual ingredients of a modern electronic management system – fuel injectors, input sensors, an Electronic Control Unit and so on – the use of accelerator position sensing and an electronic throttle actuator makes this system very different to those that came before. For the first time, the relationship between accelerator pedal position and throttle opening became adjustable – not only could this system control injection and ignition, but also the cylinder charge.

Making the advent of the ME-Motronic even more of a sea change was the underlying operating logic. Unlike other engine management systems, ME-Motronic determines how much engine torque is required in any given situation, and electronically opens the throttle blade sufficiently to allow the engine to develop that much torque. The accelerator pedal travel becomes just the driver's

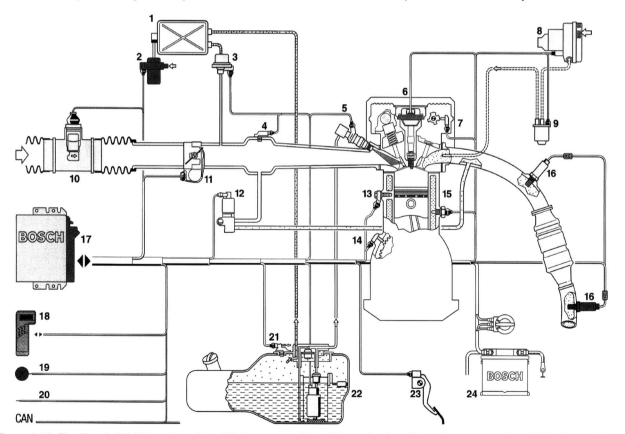

Figure 8-13: The Bosch ME-Motronic system. (1) charcoal canister, (2) check valve, (3) canister purge valve, (4) intake manifold pressure (MAP) sensor, (5) fuel rail and injector, (6) ignition coil and sparkplug, (7) camshaft position sensor, (8) air injection pump, (9) secondary air injection valve, (10) airflow meter, (11) electronic throttle valve, (12) EGR valve, (13) knock sensor, (14) crankshaft position sensor, (15) coolant temperature sensor, (16) oxygen sensors, (17) ECU, (18) diagnostics interface, (19) Check Engine lamp, (20) vehicle immobiliser, (21) tank pressure sensor, (22) in-tank pump, (23) accelerator pedal, (24) battery. (Courtesy Bosch)

WORKSHOP PRO CAR ELECTRICAL AND ELECTRONIC SYSTEMS

'torque request' input, to be weighed up against other torque requests that may be generated by the traction control system, speed limiter, engine braking torque control, and others. Additionally, at all times the engine management ECU models the engine's instantaneous torque development, adjusting the throttle opening according to the relationship between the requested and developed torque.

The system and its variants are, at the time of writing, being used in current cars. Figure 8-13 (previous page) shows an overview of the system.

Inputs and outputs

As indicated, at first glance the ME-Motronic system looks very similar to other current management systems. Figure 8-14 shows the inputs and outputs of a typical ME-Motronic system. In addition to two-way diagnostics and CAN buses (these communicate with other systems such as the automatic transmission ECU), the inputs comprise:
- Vehicle speed
- Transmission gear
- Camshaft position
- Crankshaft speed and position
- Dual oxygen sensors (located either side of the catalytic converter – 'V' engines have four sensors)
- Knock sensor
- Coolant temperature
- Intake air temperature sensor
- Battery voltage
- Intake air mass (plus frequently manifold pressure)
- Throttle position

None of these inputs is unique to this system, but the following one is:
- Accelerator pedal position

With one exception, the outputs are also very similar to other recent management systems:
- Spark plugs
- Injectors
- Instrument panel tachometer
- Fuel pump relay
- Oxygen sensor heaters
- Intake manifold runner control (ie control of the position of valves within dual tuned length manifolds, or the length of infinitely variable intake runners)
- Fuel system evaporative control, secondary air injection and exhaust gas recirculation (all emissions control approaches)

The unique addition is the:
- Electronic throttle control actuator

Given that the additional hardware comprises the accelerator pedal position sensor and electronic throttle control actuator, let's have a look at these two components in more detail.

Accelerator pedal position sensor

Two approaches are used in the design of this sensor, but they are electrically identical. Movement of the accelerator pedal manipulates two rotary potentiometers; unlike some electronic throttle engines, no back-up Bowden cable exists to connect accelerator pedal movement to the throttle blade. Two potentiometers are fitted to the sensor to allow redundancy – if one should fail, the other still lets the system operate. The outputs of the potentiometers are identical but for a voltage offset. Cars equipped with automatic transmissions do not have an additional kickdown switch in the assembly; instead a 'mechanical pressure point' is used to give the feel of a kickdown switch.

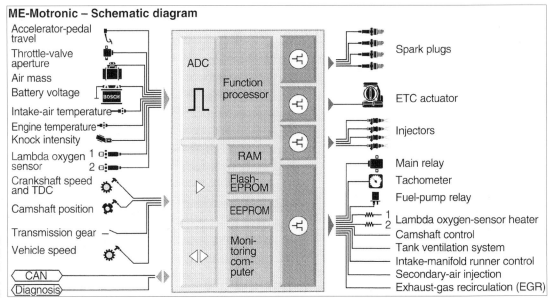

Figure 8-14: The inputs and outputs of a typical ME-Motronic system. (Courtesy Bosch)

8. ENGINE MANAGEMENT

If the accelerator position sensor fails, the lack of any mechanical connection between the accelerator and the throttle blade requires that sophisticated 'limp home' techniques are in place. Two techniques are used:

Emergency running program #1
This occurs when a single accelerator position potentiometer fails.
- Throttle position is limited to a defined value.
- In the case of implausible signals from the two potentiometers, the lower value of the two is used.
- The brakelight signal is used to indicate when idling speed should be enacted.
- The fault lamp is illuminated.

Emergency running program #2
This occurs when both accelerator position potentiometers fail.
- The engine runs only at idle speed.
- The fault lamp is illuminated.

Interestingly, if the accelerator and brake pedals are depressed together, the throttle valve is automatically closed to a defined small opening. However, if the brake is pressed and depressing of the accelerator then follows this, the torque request is enabled. I assume that the latter provision is solely for those who like to left-foot brake, with applications of power used to balance the car!

Electronic throttle control actuator

The electronic throttle valve consists of a DC motor, reduction gear drive and dual feedback angle sensors. It is again for reasons of redundancy that two potentiometers are used for angle feedback. However, unlike the accelerator position sensor, these sensors have opposite resistance characteristics to one another – ie one rises in signal voltage as the other falls.

While continuous sensing of throttle blade position does occur, the ECU recognises four key functional positions of the throttle blade:
- Lower mechanical limit stop – the valve is totally shut.
- Lower electrical limit stop – the lower limit used in normal operation. This position does not totally close the valve, thus preventing contact wear of the housing and throttle blade.
- Emergency running position – the position of the valve when it is not energised. This allows sufficient airflow for an idle speed a little higher than standard.
- Upper electrical limit stop – the blade is fully open.

The control system has a self-learning function, whereby the state of the mechanicals within the electronic throttle (eg spring tensions) is determined by the evaluation of the throttle valve's reaction speed.

As with the Accelerator Pedal Position Sensor, sophisticated limp-home techniques are available should the Electronic Throttle Control Actuator develop problems. These include:

Emergency running program #1
This occurs when an angle sensor within the throttle body fails or an implausible signal is received. Required is an intact throttle angle sensor and plausible mass airflow measurement.
- Torque increasing requests from other systems are ignored (eg from the Engine Braking Control).
- The fault lamp is illuminated.

Emergency running program #2
This occurs if the throttle valve drive fails or malfunctions; it requires that both throttle valve potentiometers recognise the Emergency Running Position of the throttle blade.
- The throttle valve drive is switched off so that the valve defaults to the small emergency running opening.
- As far as possible, ignition angle control and turbo boost control are used to execute driver torque demands.
- The fault lamp is illuminated.

Emergency running program #3
This occurs if the throttle valve position is unknown and/or if the throttle valve is not definitely known to be in the Emergency Running Position.
- The throttle valve drive is switched off so that the valve defaults to the small emergency running opening.
- The engine speed is limited to approximately 1200 rpm by fuel-injection control.
- The fault lamp is illuminated.

Torque control logic

The ME-Motronic system prioritises and coordinates torque demands in order that it can implement an overall torque control strategy. Torque requests are categorised as 'Internal' or 'External.' External torque requests include those made by the driver, cruise control system and driving dynamics systems like Automatic Stability Control. Internal torque requests are those made by the internal programming of the ECU – factors such as engine governing and idle speed control. The total requested torque is then modified by strategies such as those which take into account catalytic converter temperature or driving smoothness. Figure 8-15 (overleaf) shows the process.

In previous engine management systems, the driver – via the mechanical alteration of the throttle blade angle – exercised direct control over the mass of cylinder charge, while the management system was limited to torque reduction strategies (eg by fuel cuts) or minor torque increases through manipulation of the mass of air bypassing the throttle. However, this approach does not cope very

WORKSHOP PRO | CAR ELECTRICAL AND ELECTRONIC SYSTEMS

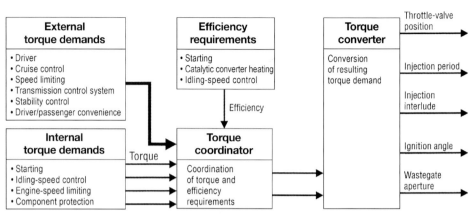

Figure 8-15: The ME-Motronic system prioritises and coordinates torque demands so that it can implement an overall torque control strategy. External torque requests include those made by the driver, cruise control system and driving dynamics systems like Automatic Stability Control. Internal torque requests are those made by the internal programming of the ECU – factors such as engine governing and idle speed control. The total requested torque is then modified by strategies such as those which take into account catalytic converter temperature or driving smoothness. (Courtesy Bosch)

well with competing and contrary torque demands that may well occur simultaneously.

The ME-Motronic system internally models the net torque development of the engine. This model takes into account losses through internal friction, pumping losses, and parasitic loads such as that of the power steering and water pumps. Internal mapping within the ECU allows optimal specifications for charge density, injection duration and ignition timing for any desired net torque value, taking into account the often-conflicting requirements of best fuel economy and emissions. These requirements dictate that the system must perform well in transients (ie sudden changes in torque), as well as when being subjected to steady-state loads. To allow good performance in both constant and transient load conditions, two different control approaches are taken.

The first control strategy is termed by Bosch the Charge Path. 'Charge' in this context refers to the mass density of air trapped in the cylinder. At a given air/fuel ratio and ignition advance, the mass of this air is directly proportional to the force generated during the combustion process. The Charge Path, controlled by the opening angle of the throttle blade (and boost pressure in a forced induction car), is used to control engine torque output in static operations. The ability of this control system to change quickly is limited by the regulating speed of the throttle actuator and the time constant of the intake manifold, which can be as high as several hundred milliseconds at low engine speeds.

The other technique used to control torque output is termed, somewhat oddly, the Crankshaft Synchronous Path. This refers to torque variations able to be rapidly created by changes in ignition timing and injection operation, with the latter used to affect the air/fuel ratio. Examples of when this approach is employed include torque reduction during automatic transmission gear changes and when Vehicle Stability Systems are operating.

Figure 8-16 puts it all together. On the far left is the driver, who (at least on the diagram!) is still given pride of place. The driver request for torque is prioritised and processed in terms of driveability functions. These include filtering and slope-limiting, dashpot (to ensure that torque changes do not occur too quickly) and anti-jerking. These functions can be calibrated to suit a wide range of applications – for example a high level of anti-jerk to suit a luxury car, or very quick throttle response to suit a sports car.

In addition to the driver's torque request, other torque variations (for example, an increase in torque to operate the air conditioner compressor, or a reduction in torque required by the load change damping system) are processed, with the final request then fed into the 'Torque to charge density conversion' box. When a torque request is made, the ECU must calculate how much fresh air mass is required to be inhaled by the engine to meet this demand. The actual mass of air that is needed will be dependent on ignition timing (eg if the engine is running relatively retarded ignition to decrease oxides of nitrogen emissions, more air will be needed because efficiency will be lower), internal engine friction, the instantaneous air/fuel ratio and other factors.

Once a mass airflow that will meet the requirements is quantified, a throttle valve opening angle is calculated. However, in all engines, the required angle will be dependent on the manifold pressure, and in forced aspirated engines, manifold pressure will be quite critical to the mass of air actually inhaled. Thus, in these engines, the turbocharger boost pressure and throttle valve opening are both specified such that the appropriate charge density required for the prescribed torque output is reached.

Calculating cylinder charge

As can be seen from the above, the accurate calculation of cylinder charge is vital if the torque modelling strategy is to be effective, and if appropriate amounts of fuel are to be accurately added to this air. Traditionally, a mass airflow

8. ENGINE MANAGEMENT

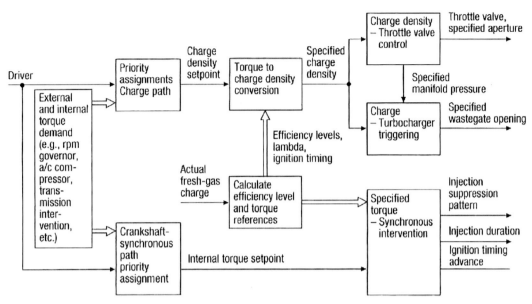

Figure 8-16: A flow chart showing the complete torque development approach. The driver request for torque is prioritised and processed in terms of driveability functions. These include filtering and slope-limiting, dashpot (to ensure that torque changes do not occur too quickly) and anti-jerking. (Courtesy Bosch)

meter positioned between the airfilter box and the throttle body has been used to measure intake airflow. However, the mechanical design of engines is now taking advantage of techniques that maximise cylinder charge in a way in which an averaged mass airflow measurement may not be able to accurately sense.

In the ME-Motronic system the available sensors are used as inputs to a charge air model, rather than being evaluated directly. The requirements for such a charge air model are:

- Accurate mass charge air determination in engines using resonant tuned and/or variable length intake manifolds, and engines using variable valve timing.
- Accurate response to Exhaust Gas Recirculation conditions.
- Calculation of required throttle valve aperture (and required turbo boost in forced induction engines).

While the engine is subjected to a constant load, mass airflow measurement is relatively accurate: ie if Xkg of air per second is passing through the airflow meter, it can be assumed that all of it is ending up in the cylinders! However, during transients, the situation is much more complex. For example, if the throttle blade is abruptly opened, the intake plenum chamber will rapidly fill with air. For an instant, this will give an inaccurately high reading from the airflow meter – the meter will indicate a higher cylinder charge than has actually had time to occur. It is only when intake manifold pressure has risen that the flow will commence into the cylinders.

As a result of this characteristic, the ME-Motronic system generally uses both manifold absolute pressure (MAP) and hot wire airflow meter (HFM) inputs. (In some cases, the MAP sensor is not fitted; further software modelling duplicates its function.) The HFM is a further development of the design used by Bosch and other management systems. Its improvements result in better accuracy; for example, it is capable of differentiating reverse flow pulses from airflow passing into the engine.

DIRECT PETROL (GASOLINE) INJECTION

As we have seen, conventional electronic fuel injected engines use injectors to add fuel to the intake ports. In a sequentially injected engine, each injector opens a short time before its associated intake valves, while in simultaneously injected engines, the injectors open all at the same time. In either case, the cloud of small fuel droplets

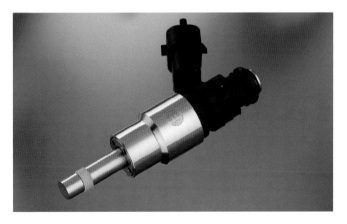

The fuel injector in a direct injection system is subjected to enormous pressures in an environment of heat and vibration. Minimum injector opening time is just 5ms and droplets are on average smaller than 20μm – only one-fifth the droplet size of traditional injectors and one-third the diameter of a human hair. (Courtesy Bosch)

WORKSHOP PRO CAR ELECTRICAL AND ELECTRONIC SYSTEMS

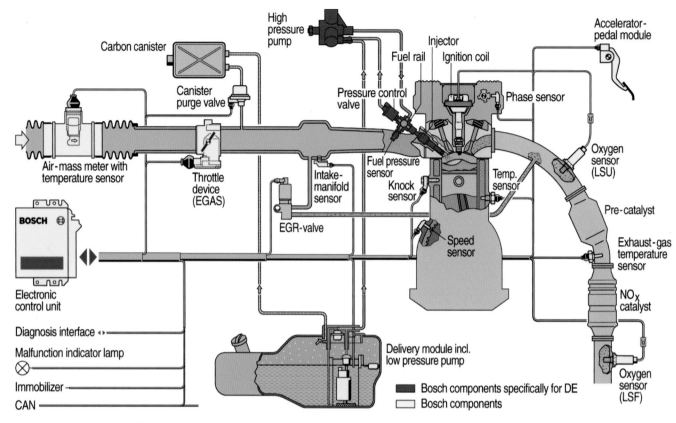

Figure 8-17: The layout of a typical Bosch direct injection system. (Courtesy Bosch)

is drawn into the engine only when the intake valves open. In contrast, Direct Petrol Injection engines squirt the fuel straight into the combustion chambers.

Like diesel engines, the air/fuel mixing occurs inside the combustion chamber, rather than in the inlet ports. Taking this approach gives far greater control over the combustion process, allowing for a variety of combustion operating modes, including those having ultra-lean air/fuel ratios. However, the degree of electronic control required to smoothly transition from one combustion operating mode to another is complex, and engine operating processes need to be monitored far more closely than is the case with conventional port fuel-injection.

System mechanicals

Figure 8-17 shows the layout of the Bosch direct injection system. Its mechanical elements differ from conventional port injection in three important ways.

First, the fuel supply system uses two fuel pumps – a conventional electrical fuel pressure pump (in the past dubbed a high-pressure pump, but now referred to in this system as a low-pressure pump) and a mechanically-driven high-pressure pump. The low-pressure pump works at pressures of 0.3-0.5MPa (43-72psi) while the high-

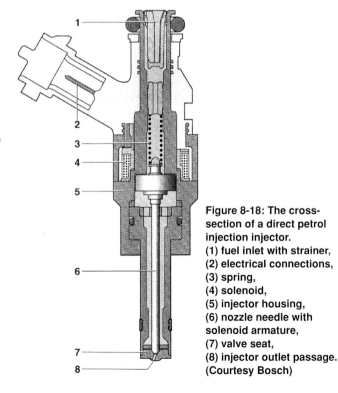

Figure 8-18: The cross-section of a direct petrol injection injector.
(1) fuel inlet with strainer,
(2) electrical connections,
(3) spring,
(4) solenoid,
(5) injector housing,
(6) nozzle needle with solenoid armature,
(7) valve seat,
(8) injector outlet passage.
(Courtesy Bosch)

106

pressure pumps boost this very substantially to 5-12MPa. (725-1740psi). The high-pressure fuel is stored in the fuel rail that feeds the injectors. The fuel rail is made sufficiently large that pressure fluctuations within it are minimised as each injector opens. The pressure of the fuel in the injector supply rail is controlled by an electronically-controlled bypass valve that can divert fuel from the high-pressure pump outlet back to its inlet. The fuel bypass valve is varied in flow by being pulse-width modulated by the ECU. A fuel pressure sensor is used to monitor fuel rail pressure.

Second, compared with a conventional port fuel-injection system, the fuel injectors must be capable of working with huge fuel pressures, and also injecting large amounts of fuel in very short periods. Figure 8-18 shows a cross-sectional view of an injector. The reason for the much-reduced time in which the injection can be completed is because all the injection must occur within a portion of the induction stroke. Port fuel injectors have two complete rotations of the crankshaft in which to inject the fuel charge – at an engine speed of 6000rpm, this corresponds to 20ms. However, in some modes, direct fuel injectors have only 5ms in which to inject the full-load fuel. The fuel requirements at idle can drop the opening time to just 0.4ms. Direct injection fuel droplets are on average smaller than 20μm – only one-fifth the droplet size of traditional injectors and one-third the diameter of a human hair.

Finally, the very lean air/fuel ratios at which direct injection systems can operate results in the production of large quantities of oxides of nitrogen (NO_x). As a result, direct injected cars require both a primary catalytic converter fitted close to the engine, and also a main catalytic converter – incorporating a NO_x accumulator – that is fitted further downstream.

Combustion modes
The really radical nature of direct fuel-injection can be seen when the different combustion modes are examined. There are at least six different ways in which combustion can take place.

1.   Stratified charge mode
At low torque outputs up to about 3000rpm, the engine is operated in stratified charge mode. In this mode, the injector adds the fuel during the compression stroke, just before the sparkplug fires. In the period between the injection finishing and the sparkplug firing, the airflow movement within the combustion chamber transports the air/fuel mixture into the vicinity of the sparkplug. This results in a portion of relatively rich air/fuel mixture surrounding the sparkplug electrode while the rest of the combustion chamber is relatively lean. The gas filling the rest of the chamber often comprises recirculated exhaust gases which results in a reduced combustion temperature

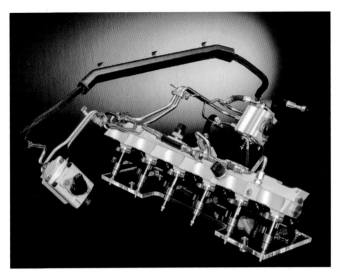

The BMW V12 direct injection injectors and fuel rail. Fuel rail pressure varies from 30-100Bar (435 – 1450psi). The injection pumps are fitted above the outlet camshafts and are driven by an additional cam. (Courtesy BMW)

and so decreased NO_x emissions. In Bosch direct injection systems, the air/fuel ratio within the whole combustion chamber can be as lean as 22:1-44:1. Mitsubishi states that total combustion chamber air/fuel ratios of 35-55:1 can be used. This can be compared with a conventional port fuel injected engine that seldom uses an air/fuel ratio leaner than 14.7:1.

2.   Homogeneous mode
Homogeneous mode is used at high torque outputs and at high engine speeds. Injection starts on the intake stroke so there is sufficient time for the air/fuel mixture to be distributed throughout the combustion chamber. In this mode, Bosch systems use an air/fuel ratio of 14.7:1 (the same as with port fuel-injection at light loads), while Mitsubishi use air/fuel ratios from 13-24:1.

3.   Homogeneous lean-burn mode
In the transition between stratified and homogeneous modes, the engine can be run with a homogeneously lean air/fuel ratio.

4.   Homogeneous stratified charge mode
Initially, this mode appears nonsensical – how can the combustion process be both homogeneous and stratified? However, what occurs is not one but two injection cycles. The initial injection occurs during the intake stroke, giving plenty of time for the fuel to mix with the air throughout the combustion chamber. Then, during the compression stroke, a second amount of fuel is injected. This leads to the creation of a rich zone around the sparkplug. The rich

Figure 8-19: Although there are at least six different modes of combustion that can occur in a direct injection engine, this diagram shows the two main modes. In Stratified Mode the injector adds the fuel during the compression stroke, just before the sparkplug fires. In the period between the injection finishing and the sparkplug firing, the airflow movement within the combustion chamber transports the air/fuel mixture towards the sparkplug. This results in a portion of relatively rich air/fuel mixture surrounding the sparkplug electrode while the rest of the combustion chamber is relatively lean. In Homogeneous Mode, injection starts on the intake stroke, so there is sufficient time for the air/fuel mixture to be distributed throughout the combustion chamber. Stratified Mode can result in air/fuel ratios as lean as an incredible 55:1. (Courtesy Bosch)

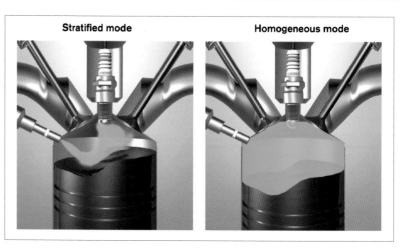

zone easily ignites, which in turn ignites the leaner air/fuel ratio within the remainder of the combustion chamber. Of the total fuel addition, approximately 75 per cent occurs during the first injection and 25 per cent in the second. The homogeneous stratified charge mode is used during the transition from stratified charge to homogeneous Modes.

In addition, there are at least two more modes – homogeneous anti-knock and stratified charge cat-heating. The first is used at full throttle, and the second to rapidly heat the catalytic converter to operating temperature. A final mode – mentioned in only some of the literature – is rich homogeneous mode, which is used to regenerate the NO_x cat. (The NO_x cat deposits oxides of nitrogen in the form of NHO_3 nitrates. When the cat is regenerated, the nitrate, together with carbon monoxide, is reduced in the exhaust to nitrogen and oxygen.)

Figure 8-19 shows the two primary combustion modes.

Electronic control systems

As was indicated earlier, the injectors must be opened against very high fuel pressures. In order that this can be achieved, a peak/hold strategy is employed whereby the opening current is very high and the 'hold' current much reduced. A dedicated triggering module is used to control the injectors, with a booster capacitor providing 50-90V to initially open the injector. Figure 8-20 shows this process.

The sensing of the mass of cylinder charge is more complex on a direct injected engine than a conventional port injected engine. This is because, at times, recirculated exhaust gas forms a major component of the total cylinder charge. As a result, two-cylinder charge sensors are used. These comprise a conventional hot-film mass airflow sensor (ie similar to a hot-wire airflow meter) and a manifold pressure sensor (MAP sensor). The flow through the airflow meter is used as an input into the calculation of the pressure within the intake manifold and this is then compared with the actual intake manifold pressure measured by the MAP sensor. The difference between the two indicates the mass flow of the recirculated exhaust gas.

As with many conventional engine management systems, direct injection requires the use of an electronically-controlled throttle. However, unlike conventional systems where the actual throttle opening more or less follows the driver's accelerator pedal torque request, in the case of direct injected engines, for much of the time the throttle is fully open – engine torque output is instead regulated by varying the fuel delivery.

Figure 8-21 shows how this occurs. During Stratified Charge Mode, the throttle is held wide-open, irrespective of the driver's accelerator pedal input. When the torque

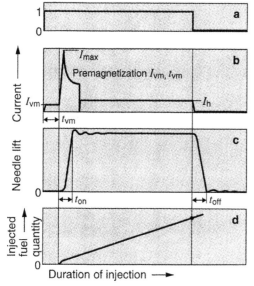

Figure 8-20: Direct injection injectors use a peak/hold system of operating. (a) is the triggering signal from the ECU, (b) is the actual injector current, (c) injector needle lift, (d) injected fuel quantity. A booster capacitor is used to provide the high opening current. (Courtesy Bosch)

8. ENGINE MANAGEMENT

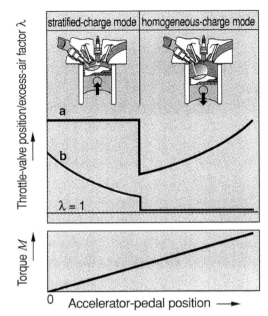

Figure 8-21: During Stratified Charge Mode the throttle is held wide-open, irrespective of the driver's accelerator pedal input. In this mode the air/fuel ratio (Bosch refer to an increased 'excess air ratio') is very lean when the torque request is low, with the air/fuel ratio gradually becoming richer as more torque is required. At a certain point, the engine changes to Homogeneous Mode. With the change in modes, the throttle valve opening becomes related to the driver's torque request and the air/fuel ratio holds a constant stoichiometric ratio (that is, 14.7:1 or Lambda = 1) across the rest of the engine load range. (Courtesy Bosch)

request is low, the air/fuel ratio is very lean (Bosch refers to this as an increased 'excess air ratio'), with the air/fuel ratio gradually becoming richer as more torque is required. At a certain point, which corresponds on an engine-specific basis to engine speed and the amount of torque required, the engine changes to Homogeneous Mode. (For simplicity, the transitional Homogeneous Lean-Burn Mode is ignored in this diagram.) With the change in modes, the throttle valve opening becomes related to the driver's torque request and the air/fuel ratio holds a constant stoichiometric air/fuel ratio (that is, 14.7:1 or Lambda = 1) across the rest of the engine load range.

The system incorporates an operating-mode co-ordinator which maps operating mode against engine speed and torque request. Figure 8-22 shows a schematic diagram of the functioning of this controller. As can be seen, a ten-stage prioritisation is used when ascertaining the required operating mode. Before the selected combustion mode starts to occur, control functions for exhaust-gas recirculation, fuel tank ventilation, charge-movement flap (ie port tumble valves or variable length intake manifold), and electronic throttle settings are initiated as required. The system waits for acknowledgement that these actions have been carried out before altering fuel-injection and ignition timing.

The advantage of having the electronic throttle valve fully open at low loads is a huge reduction in pumping losses – the engine is no longer trying to breathe through the restriction of the nearly-closed throttle. However, the downside of this is that the partial vacuum that is normally available for the brake booster will be lacking. To overcome this problem, a vacuum switch or pressure sensor monitors brake booster vacuum and if it is necessary, the combustion mode is altered so that vacuum again becomes available.

Increased efficiencies

In addition to the reduction in pumping losses occurring as a result of the throttle being wide open at low loads, during Stratified Charge Mode thermodynamic efficiencies are also increased. This is because the rich cloud of

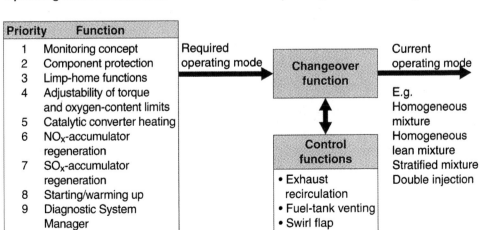

Figure 8-22: The direct injection system incorporates an operating-mode co-ordinator. As can be seen, a ten-stage prioritisation is used when ascertaining the required operating mode. Before the selected combustion mode starts to occur, control functions for exhaust-gas recirculation, fuel tank ventilation, charge-movement flap and electronic throttle settings are initiated as required. The system waits for acknowledgement that these actions have been carried out before altering fuel-injection and ignition timing to provide the appropriate combustion mode. (Courtesy Bosch)

Figure 8-23: The high-pressure pump is driven directly from the engine and develops fuel pressures as high as 12MPa (1700psi). This high-pressure fuel is regulated in pressure by a fuel pump bypass valve which is pulse width modulated by the Electronic Control Unit. The injectors are opened with a burst of high current from a capacitor that delivers up to 90V. Fuel rail pressure is monitored by a dedicated sensor. (Courtesy Bosch)

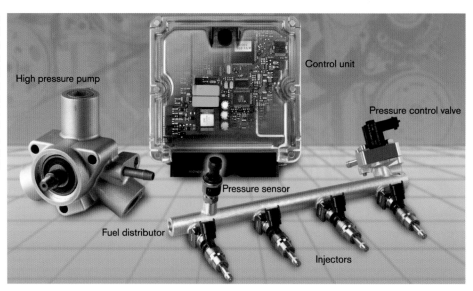

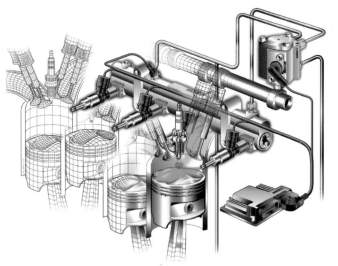

combustible air/fuel mixture around the sparkplug is thermally insulated by the layer of air and recirculated exhaust gas that surrounds it. Together with the much leaner air/fuel ratios than can be used in a conventional port injected engine, the result is a fuel efficiency improvement that can be up to 40 per cent at idle. During homogeneous mode operation, both the use of an air/fuel ratio that is never richer than 14.7:1, and the higher compression ratios normally associated with direct injection engines, result in a fuel saving of about 5 per cent.

Figure 8-24: Fuel is introduced directly into the combustion chamber by the high-pressure injectors. Depending on the operating mode, the fuel can be added during the intake stroke, during the compression stroke, or during both the intake and compression strokes. (Courtesy Bosch)

Figure 8-25: There are two main approaches to aiming the fuel spray. In Wall Guided (left), the air movement within the combustion chamber guides the area of rich mixture in the direction of the sparkplug. In Spray Guided (right), the fuel is injected directly into the vicinity of the sparkplug. The latter approach gives improved emissions and fuel economy but thermally stresses the sparkplug. (Courtesy Bosch)

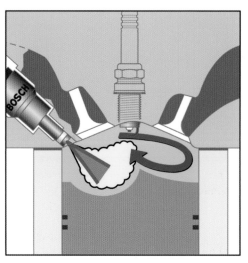

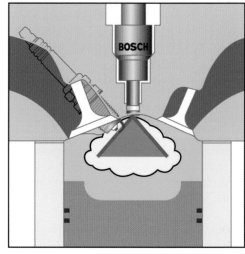

8. ENGINE MANAGEMENT

COMMON RAIL DIESEL ENGINE MANAGEMENT
Although the basic designs of petrol and diesel engines are similar (both are two or four stroke designs which use reciprocating pistons driving a crankshaft), a diesel engine does not compress its fuel/air charge and then initiate combustion by the use of a sparkplug. Instead, in a diesel engine just air is compressed. When the piston is near Top Dead Centre, the fuel is sprayed by an injector into the combustion chamber, whereupon it mixes with the hot compressed air and self-ignites.

In order that the air within the diesel combustion chamber reaches an adequate temperature for self-ignition to occur, the compression ratio needs to be much higher than found in a spark ignition engine. Compression ratios in the range of 16:1 to 24:1 are commonly used, giving forced aspirated diesel engines a compression pressure of up to 150Bar (about 2200psi). This generates temperatures of up to 900°C (1650°F). Since the ignition temperature of the most easily combustible components of diesel fuel is only 250°C (480°F), it is easy to see why the fuel burns when it is injected after the piston has risen on the compression stroke.

Diesel engines are designed to develop high torque at low engine speeds, resulting in better fuel economy. In recent years, the use of turbochargers and common rail direct injection have dramatically improved the specific torque output of diesel car engines.

Compared with petrol-powered engines (other than direct injection engines) that most often run with stoichiometric mixtures, diesels use very lean air/fuel ratios. The air/fuel ratios for diesel engines under full load are between 17:1 and 29:1, while when idling or under no load, this ratio can exceed 145:1. However, within the combustion chamber, localised air/fuel ratios vary – it is not possible to achieve a homogeneous mixing of the fuel with the air within the combustion chamber. To reduce these in-chamber air/fuel ratio variations, large numbers of very small droplets of fuel are injected. Higher fuel pressure results in better fuel atomisation, so explaining the increase in injection pressures now used.

Injection
Diesel engines are not throttled. Instead, the combustion behaviour is affected by these variables:
- Timing of start of injection
- Injection duration
- Injector discharge curve

Since the use of electronically controlled common rail injection allows these variables to be individually controlled, we'll briefly look at each.

Timing of start of injection
The timing of the injection of fuel has a major effect on emission levels, fuel consumption and combustion noise. The optimal timing of the start of injection varies with engine load. In car engines, optimal injection at no load is within the window of 2 crankshaft degrees Before Top Dead Centre (BTDC) to 4 degrees After Top Dead Centre (ATDC). At part load, this alters to 6 degrees BTDC to 4 degrees ATDC, while at full load the start of injection should occur from 6-15 degrees BTDC. The duration of combustion at full load is 40-60 degrees of crankshaft rotation.

Too early an injection initiates combustion when the piston is still rising, reducing efficiency and so increasing fuel consumption. The sharp rise in cylinder pressure also increases noise. Too late an injection reduces torque and can result in incomplete combustion, increasing the emissions of unburned hydrocarbons.

Injection duration
Unlike a conventional port fuel injected petrol engine, where the amount of fuel injected can be considered to be directly proportional to the injector opening time, a diesel injector will vary in mass flow depending on the difference between the injection and combustion chamber pressures, the density of the fuel (which is temperature dependent), and the dynamic compressibility of the fuel. The specified injector duration must therefore take these factors into account.

Discharge curve
Diesel fuel injectors do not add the fuel for a combustion cycle in one event, instead they operate in up to four different modes. The first is pre-injection, a short duration pulse which reduces combustion noise and Oxides of Nitrogen (NO_x) emissions. The bulk of the fuel is then added in the main injection phase, before the injector is turned off momentarily before then adding a post-injection amount of fuel. This post-injection reduces soot emissions. Finally, at up to 180 crankshaft degrees later, a retarded post-injection can occur. The latter acts as a reducing agent for an NO_x accumulator-type catalytic converter and/or raises the exhaust gas temperature for the regeneration of a particulate filter.

The injection amounts vary between $1mm^3$ for pre-injection to $50mm^3$ for full-load delivery. The injection duration is 1-2ms.

Common rail system overview
Unlike previous diesel fuel-injection systems – even those electronically controlled – common rail systems use, as the name suggests, a common fuel pressure rail that feeds all injectors. By separating the functions of fuel pressure generation and fuel-injection, a common rail system is able to supply fuel over a broader range of injection timing and pressure than previous systems.

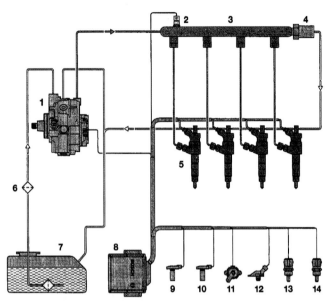

Figure 8-26: A simple common rail diesel fuel-injection system. A high pressure mechanical pump (1) pressurises the fuel which flows to the common rail (3). A fuel rail control valve (4) allows the fuel pressure to be maintained at a level set by the Electronic Control Unit (8). The common rail feeds the injectors (5). Sensor inputs to the ECU comprise fuel pressure (2), engine speed (9), camshaft position (10), accelerator pedal travel (11), boost pressure (12), intake air temperature (13) and engine coolant temperature (14). (6) and (7) are the fuel filter and fuel tank, respectively. (Courtesy Bosch)

Figure 8-26 shows a simple common rail fuel-injection system. A high pressure mechanical pump pressurises the fuel which flows to the common rail. A control valve allows the fuel pressure to be maintained at a level set by the ECU. The common rail feeds the injectors, which are electrically operated solenoid valves. Sensor inputs to the ECU comprise fuel pressure, engine speed, camshaft position, accelerator pedal travel, boost pressure (most engines are turbocharged), intake air temperature and engine coolant temperature.

More complex common rail systems use these additional sensors:
* Vehicle speed
* Exhaust temperature
* Wideband exhaust oxygen sensor
* Differential pressure sensor (to determine cat converter and/or exhaust particulate filter blockage)

Not shown on these diagrams are the glow plugs. Common rail diesels still use glow plugs, however their use is not normally required except for starting in ambient temperatures below 0°C (32°F).

Extra ECU outputs can include control of turbocharger boost pressure, exhaust gas recirculation and intake port tumble flaps.

High pressure pump

Fuel pressures of up to 1600bar (23,000psi) are generated by the high pressure pump. This pump, which is driven from the crankshaft, normally comprises a radial piston design of the type shown in Figure 8-27. The pump is lubricated by the fuel and can absorb up to 3.8kW (5hp). So that pump flow can be varied with engine load, individual pistons of the pump are able to be shut down. This is achieved by using a solenoid to hold the intake valve of that piston open. However, when a piston is deactivated, the fuel delivery pressure fluctuates to a greater extent than when all three pistons are in operation.

Pressure control valve

The fuel pressure control valve comprises a fuel-cooled solenoid valve, as shown in Figure 8-28. The valve opening is varied by its solenoid coil being pulse width modulated at a frequency of 1KHz. When the pressure control valve is not activated, its internal spring maintains a fuel pressure of about 100Bar (1450psi). When the valve is activated, the force of the electromagnet aids the spring, reducing the opening of the valve and so increasing fuel pressure. The fuel pressure control valve also acts as a mechanical pressure damper, smoothing the high frequency pressure pulses emanating from the radial piston pump when less than three pistons are activated.

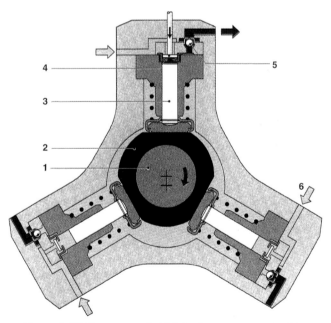

Figure 8-27: The extremely high fuel pressure required for common rail diesel injection is provided by a mechanically driven three-piston pump. (1) driveshaft, (2) drive cam, (3) pump piston, (4) intake valve, (5) outlet valve, (6) fuel inlet. (Courtesy Bosch)

8. ENGINE MANAGEMENT

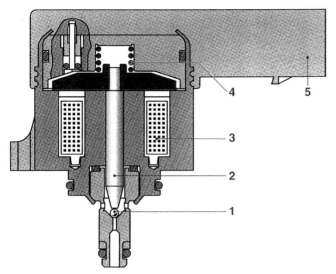

Figure 8-28: The fuel pressure regulator is electronically controlled. It comprises a fuel-cooled solenoid valve with the valve opening varied by pulse width modulation at a frequency of 1KHz. (Courtesy Bosch)

Fuel rail

The fuel rail feeds each injector. It is made sufficiently large that the internal pressure is relatively unaffected by fuel being released from the injectors. As indicated earlier, the rail is fitted with a fuel pressure sensor. To guard against dangerously high fuel pressure, a fuel pressure relief valve is also fitted.

Fuel injectors

The fuel injectors superficially look like the injectors used in conventional petrol injection systems but in fact differ significantly. Figure 8-29 shows a common rail injector. Because of the very high fuel rail pressure, the injectors use a hydraulic servo system to operate. In this design, the solenoid armature controls not the pintle but instead the movement of a small ball which regulates the flow of fuel from a valve control chamber within the injector.

When the injector is off, the ball seals the outlet from the valve control chamber. The hydraulic force acting on the end of the plunger is then greater than that acting on a shoulder located lower on the plunger, so keeping the injector closed. However, when the armature is energised, the ball is lifted and so the pressure in the valve control chamber drops. As soon as the force on the shoulder of the plunger exceeds the force on the top of the plunger, the plunger rises, lifting the pintle and allowing fuel to flow out of the injector.

The life of a common rail diesel fuel injector is certainly a hard one. Bosch estimates a commercial vehicle injector will open and close more than a billion times in its service life.

Emissions

Five major approaches are taken to reducing diesel exhaust emissions.

Design

Within the engine itself, the design of the combustion chamber, the placement of the injection nozzle and the use of small droplets all help reduce the production of emissions at their source. Accurate control of engine speed, injection mass, injection timing, pressures, temperatures and the air/fuel ratio are used to decrease emissions of oxides of nitrogen, particulates, hydrocarbons and carbon monoxide.

Exhaust gas recirculation

Exhaust gas recirculation, where a proportion of the exhaust gas is mixed with the intake charge, is also used to reduce oxides of nitrogen emissions. It does this by reducing the oxygen concentration in the combustion chamber, the amount of exhaust gas passing into the atmosphere, and the exhaust gas temperature. Recirculation rates can as high as 50 per cent.

Catalytic converter

Diesel oxidation-type catalytic converters can be used to reduce hydrocarbon and carbon monoxide emissions, converting these to water and carbon dioxide. So they rapidly reach their operating temperature, this type of catalytic converter is fitted close to the engine.

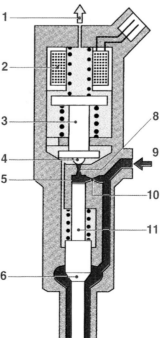

Figure 8-29: Because of the very high fuel rail pressure, the injectors use a hydraulic servo system. In this design, the solenoid armature controls not the pintle but instead the movement of a small ball (4) which regulates the flow of fuel from a valve control chamber (5) within the injector.
(1) fuel return outlet
(2) solenoid coil
(6) pressure shoulder
(7) nozzle jet
(8) outlet restrictor
(9) high pressure fuel connection
(10) inlet restrictor
(11) valve plunger.
(Courtesy Bosch)

WORKSHOP PRO CAR ELECTRICAL AND ELECTRONIC SYSTEMS

Visible soot and smoke emissions are being reduced by particulate filters like this one fitted to Mercedes cars. They can be periodically regenerated by being heated above 600°C (1100°F), a state that can be achieved by retarded injection and intake flow restriction. (Courtesy Mercedes Benz)

NO_x accumulator-type catalytic converters are also used. This type of design breaks down the NO_x by storing it over periods from 30 seconds to several minutes. The nitrogen oxides combine with metal oxides on the surface of the NO_x accumulator to form nitrates, with this process occurring when the air/fuel ratio is lean (ie there is excess oxygen). However, the storage can only be short-term and when the ability to bind nitrogen oxides decreases, the catalytic converter needs to be regenerated by having the stored NO_x released and converted into nitrogen. In order that this takes place, the engine is briefly run at a rich mixture (eg an air/fuel ratio of 13.8:1)

Detecting when regeneration needs to occur, and then when it has been fully completed, is complex. The need for regeneration can be assessed by the use of a model that calculates the quantity of stored nitrogen oxides on the basis of catalytic converter temperature. Alternatively, a specific NO_x sensor can be located downstream of the accumulator catalytic converter to detect when the efficiency of the device is decreasing. Assessing when regeneration is complete is done by either a model-based approach or an oxygen sensor located downstream of the cat; a change in signal from high oxygen to low oxygen indicates the end of the regeneration phase.

In order that the NO_x storage cat works effectively from cold, an electric exhaust gas heater can be employed.

Selective Catalytic Reduction
One of the most interesting approaches to diesel exhaust treatment is Selective Catalytic Reduction. In this approach, a reducing agent such as dilute urea solution is added to the exhaust in minutely measured quantities. A hydrolysing catalytic converter then converts the urea to ammonia, which reacts with NO_x to form nitrogen and water. This system is so effective at reducing NO_x emissions that leaner than normal air/fuel ratios can be used, resulting in improved fuel economy. The urea tank is filled at each service. AdBlue is an example of such a fluid.

Particulate filters
Exhaust particulate filters are made from porous ceramic materials. When they become full, they can be regenerated by being heated to above 600°C (1100°F). This is a higher exhaust gas temperature than is normally experienced in diesels and to achieve this, retarded injection and intake flow restriction can be used to increase the temperature of the exhaust gas.

Engine management
The engine management system in a diesel common rail engine needs to provide:
- Very high fuel-injection pressures (up to 2000bar – 29,000psi)
- Variation in injected fuel quantity, intake manifold pressure and start of injection to suit engine operating conditions
- Pre-injection and post-injection
- Temperature-dependent rich air/fuel ratio for starting
- Idle speed control independent of engine load
- Exhaust gas recirculation
- Long-term precision

As with current petrol engine management systems, the driver no longer has direct control over the injected fuel quantity. Instead, the movement of the accelerator pedal is treated as a torque request and the actual amount of fuel injected in response is dependent on the engine operating status, engine temperature, the likely effect on exhaust emissions, and the intervention by other car systems (eg traction control).

Figure 8-30 shows an overview of the inputs, outputs and internal processes in the Bosch common rail management system.

Starting
The injected fuel quantity and start of injection timing required for starting are primarily determined by engine coolant temperature and cranking speed. Special strategies are employed for very cold weather starting, especially at high altitudes. In these conditions, the turbocharger operation may be suspended as its torque demand – although small – may be sufficiently great as to prevent the car from moving off.

Driving
In normal driving, the injected fuel quantity is determined primarily by the accelerator pedal sensor position, engine speed, fuel and intake air temperatures. However, many other maps of data also influence the fuel-injection quantity

8. ENGINE MANAGEMENT

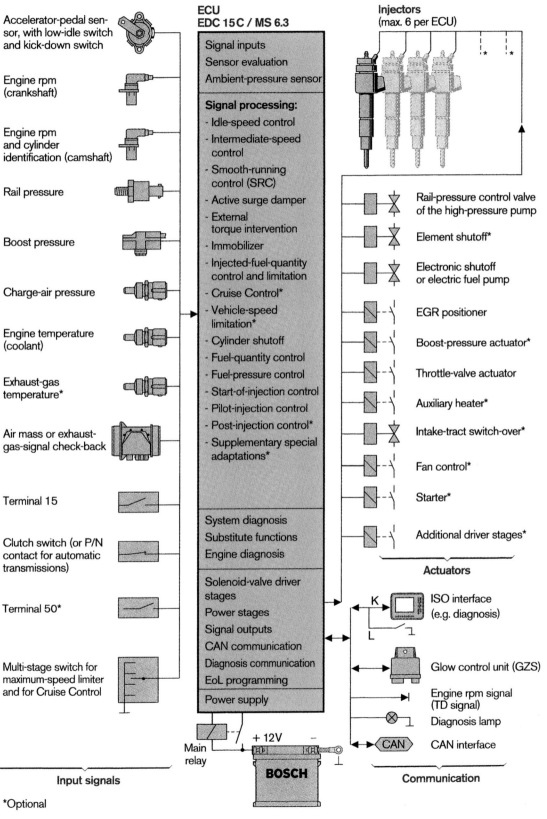

Figure 8-30: An overview of a common rail diesel engine management system. The input signals to the ECU are on the left and include accelerator pedal position, intake mass airflow, fuel rail pressure and engine speed. Not shown here but also often included is a wideband exhaust gas oxygen sensor. The outputs (right) include the control of the fuel injectors, exhaust gas recirculation (EGR) and fuel rail pressure. Inside the ECU (middle) control strategies are implemented for idle speed, smooth running control, quantity of fuel injected, starting point of injection, and many others. (Courtesy Bosch)

The Electronic Control Unit for a four-cylinder common rail BMW diesel. (Courtesy BMW)

Rev limiter

Unlike a petrol engine management system which usually cuts fuel abruptly when the rev limit is reached, a diesel engine management system progressively reduces the quantity of fuel injected as the engine speed exceeds the rpm at which peak power is developed. By the time maximum permitted engine speed has been reached, the quantity of fuel injected has dropped to zero.

Surge damping

Sudden changes in engine torque output can result in oscillations in the vehicle's driveline. This is perceived by the vehicle occupants as unpleasant surges in acceleration. Active Surge Damping reduces the likelihood of these oscillations occurring. Two approaches can be taken. In the first, any sudden movements of the accelerator pedal are filtered out, while in the second, the ECU detects that surging is occurring and actively counteracts it by increasing the injected fuel quantity when the engine speed drops and decreasing it when the speed increases. Figure 8-32 shows this process.

actually used. These include strategies that limit emissions, smoke production, mechanical overloading and thermal overloading (including measured or modelled temperatures of the exhaust gas, coolant, oil, turbocharger and injectors). Start of injection control is mapped as a function of engine speed, injected fuel quantity, coolant temperature and ambient pressure. Figure 8-31 shows example data maps for start of injection and smoke control.

Smooth running control

Because of mechanical differences from cylinder to cylinder, the development of torque by each cylinder is not identical. This difference can result in rough running and increased emissions. To counteract this, Smooth Running Control uses the fluctuation in engine speed to detect output torque variations. Specifically, the system compares the engine speed immediately after a cylinder's injection with the average engine speed. If the speed has dropped, the fuel-injection quantity for that cylinder is increased. If the engine speed is above the mean, the fuel-injection quantity for that cylinder is decreased. Figure 8-33 shows this process.

Idle speed control

The set idle speed depends on engine coolant temperature, battery voltage and operation of the air conditioner. Idle speed is a closed loop function where the ECU monitors actual engine speed and continues to adjust fuel quantity until the desired speed is achieved.

Figure 8-31: Along with many other variables, three dimensional ECU maps are used for both injection start timing and smoke limitation. (Courtesy Bosch)

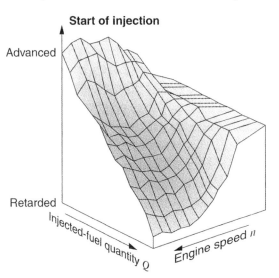

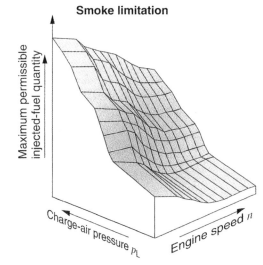

8. ENGINE MANAGEMENT

Closed loop oxygen sensor control

As with petrol management systems, diesel management system use oxygen sensor closed loop control. However, in diesel systems a wideband oxygen sensor is used that is capable of measuring air/fuel ratios as lean as 60:1. This Universal Lambda Sensor (abbreviation in German: LSU) comprises a combination of a Nernst concentration cell and an oxygen pump cell.

Because the LSU signal output is a function of exhaust gas oxygen concentration and exhaust gas pressure, the sensor output is compensated for variations in exhaust gas pressure. The LSU sensor output also changes over time and to compensate for this, when the engine is in over-run conditions, comparison is made between the measured oxygen concentration of the exhaust gas and the expected output of the sensor if it were sensing fresh air. Any difference is applied as a learned correction value.

Closed loop oxygen control is used for short- and long-term adaptation learning of the injected fuel quantity. This is especially important in limiting smoke output, where the measured exhaust gas oxygen is compared with a target value on a smoke limitation map. Oxygen sensor feedback is also used to determine whether the target exhaust gas recirculation is being achieved.

Fuel pressure and flow control

The pressure in the common rail is regulated by closed loop control. A pressure sensor on the rail monitors real time fuel pressure and the ECU maintains it as the desired level by pulse width modulating the fuel pressure control valve. At high engine speeds but low fuel demand, the ECU deactivates one of the pistons in the high-pressure pump. This reduces fuel heating in addition to decreasing the mechanical power drawn by the pump.

Other management system outputs

In addition to the control of the fuel injectors, the diesel engine management system can control:
* Glow plugs for sub-zero starting conditions
* Glow plugs that heat the coolant, providing adequate cabin heating in cold climates
* Switchable intake manifolds, where at low loads air is forced through turbulence ducts to provide better in-cylinder swirl
* Turbocharger boost pressure control
* Switching of radiator fans

Injector operation

The triggering of the injector can be divided into five phases:
* In the first phase, the injector is opened rapidly by the supply of high current from a 100V booster capacitor.

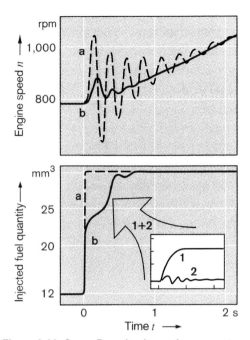

Figure 8-32: Surge Damping is used to prevent unwanted oscillations in acceleration. The top diagram shows the change in accelerations without surge damping (a) and with it (b). This alteration in car behaviour can be achieved in two ways. The lower diagram shows (1) the effect of electronic filtering of the accelerator pedal travel sensor output signal, and (2) the active correction of surge by increasing the injected fuel quantity when the engine speed drops and decreasing it when the speed increases. (Courtesy Bosch)

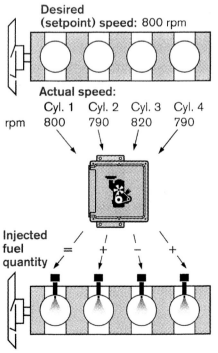

Figure 8-33: Smooth Running Control addresses the fact that the torque output of each cylinder is not identical. To counteract this, the system compares the engine speed immediately after a cylinder's injection with the average engine speed (in this case 800rpm). If the speed has dropped, the fuel-injection quantity for that cylinder is increased. If the engine speed is above the mean, the fuel-injection quantity for that cylinder is decreased. (Courtesy Bosch)

117

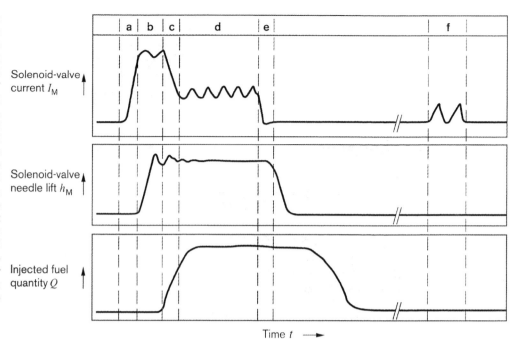

Figure 8-34: This diagram shows the relationship between solenoid valve (ie injector) current, solenoid valve needle lift and injected fuel quantity. At (a) the injector is opened with a rapidly rising (but controlled) rush of current, at (c) the current is decreased but is still sufficient to hold the injector open, at (e) the current is switched off and the injector closes. Between actual injector events (f), a sawtooth waveform is applied to the closed injector. The current used is insufficient to open the injector and the generated inductive spikes are used to further recharge the booster capacitors until they reach 100V. (Courtesy Bosch)

- Peak current is limited to 20A and the rate of current increase is controlled to allow consistent injector opening times.
- The second phase is termed 'pick-up current.' In this phase, the current supply for the injector switches from the capacitor to the battery. In this phase, peak current continues to be limited to 20A.
- A 12A pulse width modulated holding current is then used to maintain the injector in its open state. The inductive spike generated by the reduction in current through the injector in the change from 'pick-up' to 'holding' phases is routed to the booster capacitor, so starting its recharge process.
- When the injector is switched off, the inductive spike is again routed to the booster capacitor.
- Between actual injector events, a sawtooth waveform is applied to the closed injector. The current used is insufficient to open the injector and the generated inductive spikes are used to further recharge the booster capacitors until they reach 100V.

Figure 8-34 shows the relationship between injector current, needle lift and fuel flow.

OTHER ENGINE MANAGEMENT FUNCTIONS

The systems described above cover a vast range of cars, from the cheapest to the most expensive. But one thing is a constant in automotive technology – and that's that systems are always evolving. In this section, I'd like to briefly describe some of the additional approaches, sensors, and systems that I have not yet mentioned.

Variable intake manifolds

Variable intake systems change the length of the intake manifold runner or the volume of the plenum chamber. This allows the intake to have more than one tuned rpm – giving better cylinder filling at both peak torque and peak power, for example. The changeover is normally performed as a single step – the intake system is either in one configuration or the other.

The intake system can be variably tuned in a number of ways, including (especially on six-cylinder engines) connecting twin plenums at high rpm but having them remain separate smaller tuned volumes at lower revs. The introduction of a second plenum into the system at a particular rpm is another approach taken. However, the most common method is to have the induction air pass through long runners at low revs and then swap to short runners at high rpm. Note that this doesn't mean that the long runners need to be positively closed – opening parallel short runners is sufficient to change the effective tuned length of the intake system.

The changeover is normally performed by a solenoid valve which directs engine vacuum to a mechanical actuator that opens or closes the internal manifold changeover valves. Figure 8-35 shows such a system. The changeover point can be based on engine rpm (this is most common), engine load, or a combination of both.

Variable valve timing

Variable valve timing systems alter the timing and/or lift of the valves. Initially, most variable camshaft timing was on

8. ENGINE MANAGEMENT

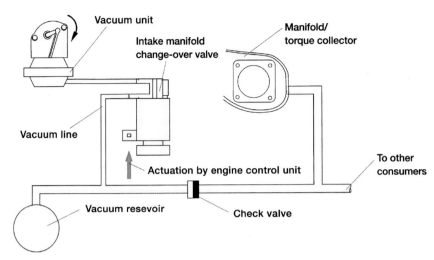

Figure 8-35: A vacuum control system for a two-length intake manifold. The ECU triggers the intake manifold change-over solenoid which feeds vacuum to an actuator that causes the intake manifold runner effective length to change. (Courtesy Volkswagen)

only one of the two camshafts and the camshaft timing was varied in a single step. That is, when the engine reached a certain rpm and/or load, the ECU moved the camshaft timing – so one cam was either in the advanced or retarded position. Depending on the engine and manufacturer, that variable cam could be either the intake or exhaust cam.

Steplessly variable cam timing was then introduced. This allows lots of 'in between' camshaft timing positions to be used, giving a far better result than single-step cam timing variation. Steplessly variable cam timing was initially used on just one camshaft, but many manufacturers now using steplessly vary cam timing on both the intake and exhaust camshafts. Figure 8-36 shows the approach taken in a Toyota VVTi system that steplessly varies only intake cam timing.

Systems that additionally vary the valve lift as well as cam timing are also employed. Honda's VTEC system is probably the best known older system of this type of single-step system. BMW has a design where the intake valve lift (as well as the exhaust and intake valve timing) are all able to be varied steplessly.

The techniques used to alter the camshaft timing and/or lift also vary. However, where the camshaft timing alters in one step, an on/off signal from the ECU is used to activate a solenoid that feeds oil pressure to the mechanism, causing the change to take place. Where camshaft timing varies steplessly, a pulsed solenoid is used to allow the cam phasers to vary in their position. Camshaft timing can be varied according to input signals including engine rpm, throttle position, engine coolant temperature and intake airflow.

Turbo boost control

All turbocharged cars of the last 30 years use electronic boost control. The system of wastegate control that is used by these cars is actually based on the older approach of using a wastegate moved by a sprung diaphragm. When boost pushing against the diaphragm overcomes the spring pressure, the diaphragm is deflected, in turn moving a lever that opens the wastegate and allows exhaust gases to bypass the turbo. This prevents the turbo from rotating any faster and so limits the peak boost able to

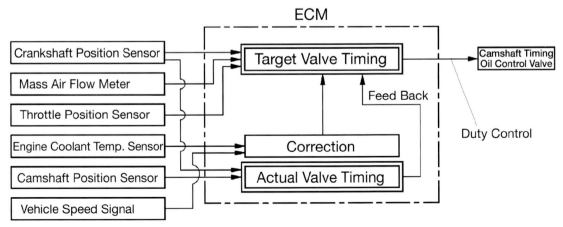

Figure 8-36: This system given stepless variation of the timing of the inlet camshaft. Note how the system can self-correct any errors that develop in timing through wear. In this system, a variable duty cycle is fed to the actuator. (Courtesy Toyota)

be developed. Electronic control adds a variable duty cycle solenoid that bleeds air from the wastegate hose, so altering the pressure that the wastegate actuator sees. Some cars also now use variable vane turbos where the geometry of the turbo itself is altered to control boost pressure.

Most turbo boost control systems are closed loop. In this approach, the boost level is monitored by a manifold pressure sensor which is able to adjust the control system to give the desired boost level, even at different altitudes and temperatures. Boost control is integrated into the engine management ECU so the system has available all of the normal engine management inputs, including intake air temperature, knock sensor activity and MAP.

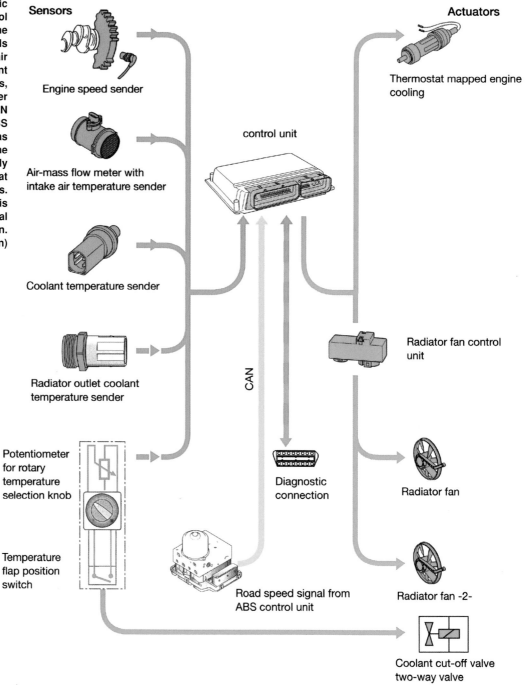

Figure 8-37: The electronic cooling system control used by Volkswagen. The ECU uses input signals of engine speed, air mass flow, two coolant temperature sensors, and the driver's heater request. Via a CAN connection from the ABS control unit, it also has access to road speed. The outputs are the electrically heated thermostat and two radiator fans. Self-diagnosis is available via the normal diagnostic connection. (Courtesy Volkswagen)

8. ENGINE MANAGEMENT

Thermal management

An increasing number of cars use active thermal control of the coolant temperature. That is, rather than using just a wax pellet thermostat, they control coolant temperature by electronic control. Advantages of active thermal management include faster warm-up, lower fuel consumption in part-throttle driving conditions and reduced emissions. In addition, some cars lower the coolant temperature at times of high load, so boosting power.

One approach is to use a thermostat that has an added electric heating element. When electrically unheated, the thermostat has a normal opening temperature of 110°C (230°F). By operating the heating element, the control system is able to open the thermostat at a temperature lower than this, as required.

Figure 8-37 shows a schematic overview of the electronic control system used by Volkswagen. The ECU uses input signals of engine speed, air mass flow, two coolant temperature sensors, and the driver's heater request. Via a CAN connection from the ABS control unit, it also has access to road speed. The outputs are the electrically heated thermostat and two radiator fans. Self-diagnosis is available via the normal diagnostic connection.

The Porsche 991 and 981 models of the Boxster and 911 use the following thermal management strategy:

1. Warm-up
- Measures aimed at reducing fuel consumption
  - Specific channelling of available heat flows to operating fluids and components that are relevant to consumption
  - Reduces heat losses to a minimum, accelerates warm-up
- Increased heating comfort
  - Specific channelling of the available heat flows into the cabin

2. At operating temperature
- Reduces consumption by increasing the coolant temperature to 105°C (221°F)

3. Cooling
- Optimises performance by reducing the coolant temperature to 85°C (185°F)

Note that in the Porsches, the coolant temperature gauge on the instrument panel shows 90°C (194°F) at normal operating temperatures that may in fact vary from 85°-105°C (185°-221°F)!

Some turbocharged cars can switch on the radiator fans (and, if fitted, also an ancillary electric coolant pump), even with the ignition switch off. This 'run-on' cooling mode is based on engine coolant temperature and fuel consumption of the last driving cycle (ie harder driving is likely to have resulted in a hotter engine). The electric cooling pump operation is operated to cool the turbo bearings and reduce oil coking.

Auto stop

When they were first introduced, one of the major benefits of hybrid cars in city conditions was their ability to turn off the internal combustion engine when the car was stationary, thus decreasing fuel consumption in stop-start urban conditions. Seeing the effectiveness of this, manufacturers started to apply this technology to conventional cars as well, simply beefing-up the starting and charging systems to cope with the more frequent engine starting then required.

One approach is to switch off the engine 1-2 seconds after the vehicle has come to a stop while being braked. This mode is enabled if:
- The vehicle is at a standstill with the brake pedal depressed. The gear selector lever must be in one of the selector lever positions D, N, P or in the manually selected driving position 1 or 2.
- The driver is detected, ie the seatbelt on the driver's side is fastened, the driver's door is closed and the brake is depressed.
- The bonnet (hood) is closed.
- The engine, battery and transmission have reached operating temperature
- The vehicle has been driven for at least 1.5 seconds and at a speed of more than 2 km/h since the last automatic engine stop.
- Road gradient doesn't exceed 10%.
- The driver-selectable Sport mode is not active, and the stability control system has not been driver deactivated.
- Auto Start-Stop function is not deactivated by the driver.
- The air-conditioning is not working at maximum, and defrost has not been selected.
- The rear foglight is not switched on.
- No trailer has been detected.

The car will restart if:
- The vehicle moves.
- Reverse gear or sport mode are engaged.
- There is a shift to a gear that is not permitted in Start-Stop mode.
- Releasing the footbrake or pressing the accelerator.
- If activation occurs of sport mode, stability control off mode, air conditioner maximum or defrost.
- Falling brake pressure.
- Energy level falls below the maximum permitted energy that can be tapped from the vehicle battery for each engine stop.

In the Porsche Cayenne, there are seven different bus systems associated with the Start Stop function, in addition to the Start Stop ECU itself. In addition, there are 17 different components (from the outside temperature sensor to the brake vacuum sensor) involved.

WORKSHOP PRO CAR ELECTRICAL AND ELECTRONIC SYSTEMS

9. OTHER CAR ELECTRONIC SYSTEMS

Chapter 9

Other car electronic systems

- Taking a systems approach to car electronics
- ABS
- Electronic Stability Control
- Electronically-controlled power steering
- Climate control
- Automatic transmission control

WORKSHOP PRO: CAR ELECTRICAL AND ELECTRONIC SYSTEMS

In the last chapter, I explored some indicative engine management systems in detail. I looked at the inputs and outputs, and explained some of the control approaches being taken by ECUs. I'd like now to look at some other car electronic systems, but I want to introduce a slightly different approach. In this approach I'll leave the ECU largely as a 'black box,' a device that makes decisions based on the input signals being fed to it. As a result of those decisions, the ECU then triggers certain outcomes – eg turning on a motor to move a vent in a climate control system, or changing the degree of assistance in electric power steering.

Taking this 'black box' – or systems – approach allows you to quickly get your head around most car electronics, even when you're not initially familiar with the particular system. Rather than seeing the system as a bewilderingly large and complex behemoth, you can immediately ask yourself: *what are the inputs, and what are the outputs?* Once those are ascertained, you can consider the functionality of the system (ie what it has to do) and then make an 'educated guess' as to the *control logic* the ECU utilises. With this level of understanding, diagnosing faults becomes far easier.

We'll call this way of doings things the 'inputs/outputs/logic' systems approach.

EXAMPLE SYSTEM: AIR SUSPENSION

As an example of this 'systems' approach, I want to describe a car system with which most of you will be *unfamiliar*: electronically-controlled air suspension. A very similar design of air suspension system has been fitted to Volkswagen, Audi, Mercedes and Range Rover vehicles. These systems use a compressor, air tank, solenoid valve block and four air springs. They maintain a constant ride height irrespective of load, and some systems do tricky things like automatically lowering the car at speed, or setting a lower ride height when the car is stopped (so making getting in and out easier).

Figure 9-1 shows an overview of an air suspension system fitted to an Audi.

So, following our 'inputs/outputs/logic' approach, what are the inputs to the air suspension system? These comprise:
1. Ride height front-left
2. Ride height front-right
3. Ride height rear-left
4. Ride height rear-right

This makes sense; if the system is to maintain a constant ride height, irrespective of load, the ECU will need to know the ride height of each corner – otherwise, how will it know if a spring needs more or less air? The system will also need to know the pressure of the air in the storage tank, so we have another input:

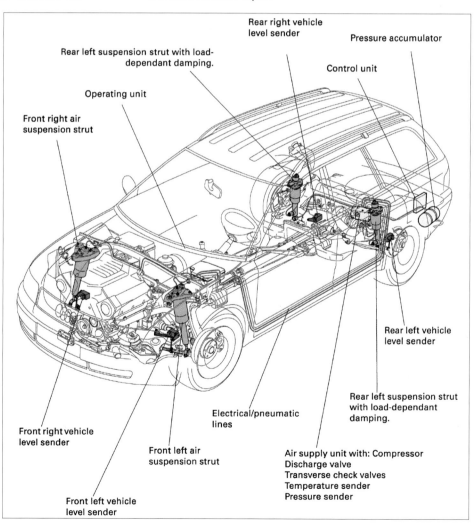

Figure 9-1: An overview of an Audi air suspension system. While it can initially look overwhelming, if we break the system down into inputs, outputs and ECU control logic, it becomes much easier to understand. (Courtesy Audi)

9. OTHER CAR ELECTRONIC SYSTEMS

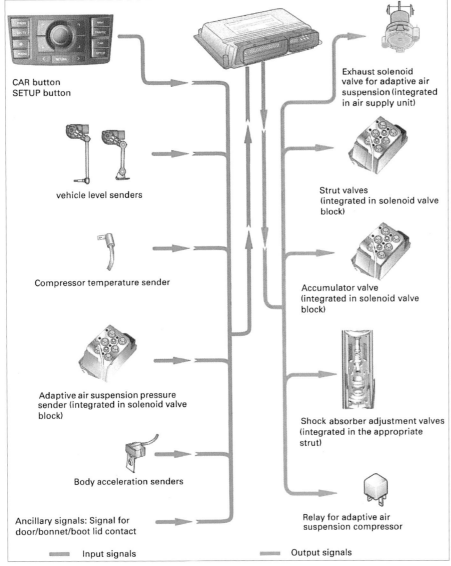

Figure 9-2: This air suspension system incorporates variable damping. Using the 'systems' approach, can you work out why this air suspension system (unlike the one described in the main text) has body acceleration senders providing additional inputs to the ECU? (Courtesy Audi)

5. Tank pressure

Some of these systems also use a temperature sensor on the head of the compressor. Again, this makes sense – if the ambient temperature is very hot and the compressor is running a lot, the compressor could overheat. So we add another input:

6. Compressor temperature

Now we don't know for sure, but it would be very likely that the air suspension system is connected (eg by CAN bus) to other systems – the Electronic Stability Control, for example. Again, this makes sense – if the Stability Control is operating because the car is nearly out of control, it won't be a good time to adjust ride height! And remember how some air suspension systems lower the car at speed or when stopped? Road speed is therefore another input

likely to be derived from the CAN bus. So we can add:

7. Bus signals from other controllers

One of the advantages of an air suspension system is that the driver can select different ride heights. This is done by a dashboard control – so another input:

8. Driver ride height selection

So we have eight inputs, with one of those inputs – the CAN bus – having the potential to carry a lot of information.

Now, what about the outputs? The system needs to be able to control the ride height of each corner separately, and it does this via four solenoid valves that allow airflow to and from the air springs. So:

1. Solenoid front-left
2. Solenoid front-right
3. Solenoid rear-left
4. Solenoid rear-right

The compressor will also need to be operated when tank pressure is low (remember, the ECU is sensing tank pressure) and so we have a relay output for the compressor:

5. Compressor relay

The driver will also need to know what is happening (especially when they are manually changing suspension height) and so there will be an output for a dash indicator light. And of course the system will also have some diagnostics:

6. Dash light
7. Diagnostics

The above output descriptions are ones that you could make almost through guesswork, but in fact a further two solenoids are used. There is one that allows the system to access the pressure tank, and another that is used to discharge air from the air springs.

8. Tank solenoid
9. Discharge solenoid

All right, now what about guessing the ECU logic? Based on the inputs and outputs, and thinking what the system

125

is supposed to achieve, we can surmise that the ECU performs these functions:

1. Monitor suspension height of each corner and add or remove air appropriately to maintain the selected ride height.
2. Turn on compressor when tank pressure is low; switch off the compressor when tank pressure reaches correct pressure.
3. Monitor driver input and alter suspension height as requested.
4. Monitor CAN bus signals and inhibit suspension height changes appropriately (eg when Stability Control or ABS are operating).
5. Monitor CAN bus signals and make suspension height changes appropriately (eg lowering car at high speed or when stopped).
6. Turn off compressor if its temperature is excessive.
7. Activate a warning light and log codes if faults are detected.

Now with that level of understanding, we can much more easily understand potential fault conditions and their symptoms.

For example, if a fault code indicates that the compressor is overheating, we can consider *why* it might be overheating. If the weather hasn't suddenly got very hot, it might be because the compressor is running a lot. The compressor runs when the tank pressure is low, so why might the tank pressure be low? It could be because the compressor is no longer performing very well (eg because its piston ring is worn) and it has to work much longer to reach the required tank pressure. However, it could also be because the system has a leak – so not the fault of the compressor at all!

Another example: the car is down on one corner and the fault code indicates a problem with the suspension height sensor for that corner. It could be that the air spring has a leak, but the combination of the two characteristics (down on that corner *and* the fault code for the height sensor) suggests that the height sensor and its wiring should be checked first.

A final example – the car no longer lowers itself when stopped. You could blame the ECU, but it would be a better starting point to suspect that the ECU is missing an input it needs in order to work out *when* to lower the car. You do a fault scan, and yes; a CAN bus error shows.

It's easy to encounter a new car electronic system and feel completely overwhelmed, but the 'inputs/outputs/ECU logic' approach allows you to quickly sort the wood from the trees.

ABS

Anti-lock brakes (ABS) work on the principle that the retardation a tyre is able to generate is greatest just prior to the wheel locking. This idea needs some explanation as the concept is not intuitive.

If the tyre on a moving car is locked, the car won't be

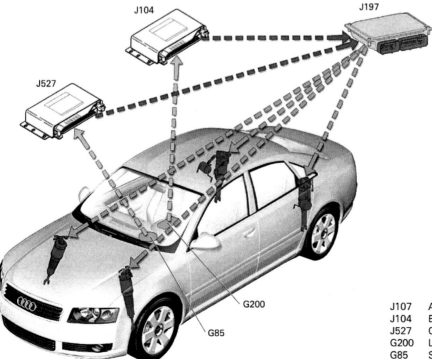

Figure 9-3: This diagram shows the other control units that the air suspension controller (J107) communicates with. Again using our 'systems' approach, can you suggest reasons why each of these other controllers would communicate with the air suspension controller? (Courtesy Audi)

J107	Adaptive air suspension control unit
J104	ESP control unit
J527	Control unit for steering column electronics
G200	Lateral acceleration sender
G85	Steering angle sender

9. OTHER CAR ELECTRONIC SYSTEMS

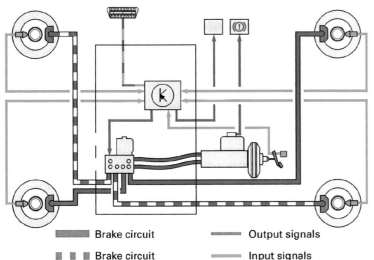

Figure 9-4: Schematic diagram of a simple ABS. The inputs are from the four wheel speed sensors and the brake pedal. The outputs are to the hydraulic control unit and dashboard warning light. The diagnostic connector has a two-way information flow. (Courtesy Volkswagen)

▬ Brake circuit   ▬ Output signals
▪ ▪ ▪ Brake circuit   ══ Input signals

slowed as fast as possible. Why? Because in this situation, the car is skimming along on a pool of molten rubber and, as a result, the wheels simply can't get a grip on the road. So a locked wheel isn't much good at stopping the car, while a wheel turning freely with no braking effort at all being applied obviously does nothing to slow the vehicle – so something in between these extremes is needed.

The amount of wheel slowing needed to get maximum stopping power is called the 'slip ratio.' A 100 per cent slip ratio means that the wheel is locked, while at zero per cent it's turning freely. The slip ratio needed to get max braking performance isn't fixed – if it were, it would all be easy. It depends on the road surface, the tyre compound and tread, road temperature and so on. Generally, a slip ratio between 8-30 per cent works best; meaning that, under braking, the tyre is rotating at up to a third slower than the road speed.

The major components of an electronic ABS system are the Electronic Control Unit (ECU), wheel speed sensors, and the Hydraulic Control Unit (HCU). In short, the ECU monitors the speed of each of the wheels, and then instructs the HCU to momentarily decrease brake fluid pressure on any wheel which is showing signs of locking.

Most systems use three or four input speed sensors, with each of the front and rear wheels monitored in four-sensor systems, and the fronts and the rear axle (or one rear wheel) monitored in three-sensor systems. The wheel speed sensors consist of a coil and a magnet, which get excited when a toothed wheel (attached to the hub or part of the brake disc) moves past. (The sensor design is similar to crankshaft position sensors used in some engine management systems.) An output pulse is generated, with the output frequency proportional to the wheel speed. The ECU can therefore sense the rotational speed of each wheel equipped with a sensor. Figure 9-4 shows a simple ABS.

In addition to the wheel speed sensors, some ABS units use a longitudinal acceleration sensor, which allows the ECU to take note of the actual amount of deceleration being experienced. One of the early problems found with ABS was on snow and dirt, where the car could actually stop faster if the wheels were locked. This is because the locked wheels gouge their way down to a firmer surface beneath, and also because a dam of snow or gravel builds

After 14 years of development, Bosch ABS was first released in 1978. It was fitted as optional equipment, at first in the Mercedes-Benz 'S'-class cars and shortly afterwards in the BMW 7-series. (Courtesy Bosch)

The classic photo of the Mercedes S Class of 1978 – the car on the left without ABS and the car on the right with the system. Note how the car on the right was able to steer around the obstacle, even under full braking. (Courtesy Mercedes Benz)

WORKSHOP PRO CAR ELECTRICAL AND ELECTRONIC SYSTEMS

up in front of the locked wheels and helps slow the car. The use of a longitudinal acceleration sensor allows the ECU to sense the actual deceleration that is occurring, and so modify its behaviour if the car is not stopping as quickly as it should.

ELECTRONIC STABILITY CONTROL

Along with ABS, Electronic Stability Control (ESC) represents the greatest safety advance in car electronic systems that has ever occurred.

ESC uses speed inputs from each wheel, a steering angle sensor, a lateral/longitudinal acceleration sensor and a yaw sensor. (Yaw is the rotation of the car around a vertical, central axis.) The main ESC outputs are to control electronic throttle opening and brake individual wheels via the ABS module.

When a vehicle is understeering (the front sliding wide), braking of the inside rear wheel will reduce the amount of understeer that occurs. For example, if the car is understeering around a right-hand bend, braking the right-hand rear wheel, while the other wheels continue to turn at their normal rate, causes the car to pivot around to the right. That's where you want the car to go when you're going around a right-hand bend, so the understeer has been reduced.

When the vehicle is oversteering (the rear sliding out), the outside front wheel needs to be braked to near lock-up to correct the slide. If the car is oversteering around a right-hand bend, this would be the front left-hand wheel. If you picture this wheel nearly stopped, but the others continuing at normal speed, you can imagine that the car attempts to pivot around to the left, reducing the amount of oversteer. Figure 9-5 shows this.

In addition to braking individual wheels, during ESC operation most cars reduce engine torque by closing the throttle, retarding the ignition timing or altering camshaft timing.

The ECU logic that ESC uses is to compare the steering wheel angle with the actual yaw rate (and/or lateral acceleration) of the car. That is, if the driver has, for example, 10° of left-hand steering lock applied at a given speed, the car should be yawing at a rate that reflects that speed and steering angle. If the car is yawing to a greater

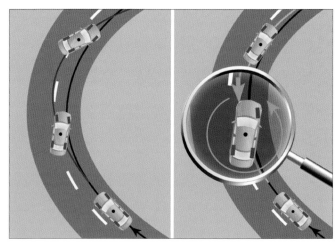

Figure 9-5: How Electronic Stability Control (ESC) works. The car on the left, without ESC, follows the path shown by the red line. Note how the driver is using opposite lock to control the oversteer. The car on the right, with ESC, much better follows the correct path. Within the magnifying glass can be seen the front-left wheel being individually braked to correct the oversteer slide. (Courtesy Mercedes Benz)

degree than it should, the car is oversteering. If the car is yawing less than it should, the car is understeering. (Incidentally, this means that in cars with ESC, you should always keep steering in the direction you want to go!)

Figure 9-6 shows an ESC system where other functions (such as ABS) are integrated into the controller.

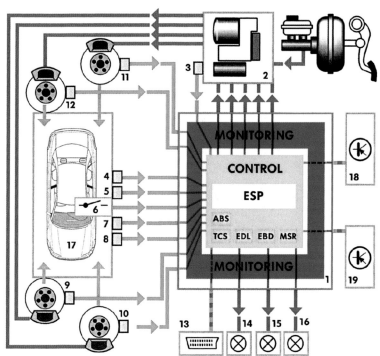

Figure 9-6: Electronic Stability Control. (1) ABS control unit with ESP and other control functions, (2) hydraulic control unit, (3) brake pressure sensor, (4) lateral acceleration sender, (5) yaw rate sender, (6) dashboard button, (7) steering angle sensor, (8) brakelight switch, (9-12) wheel speed sensors, (13) diagnosis connector, (14 – 16) warning lights. (Courtesy Audi)

9. OTHER CAR ELECTRONIC SYSTEMS

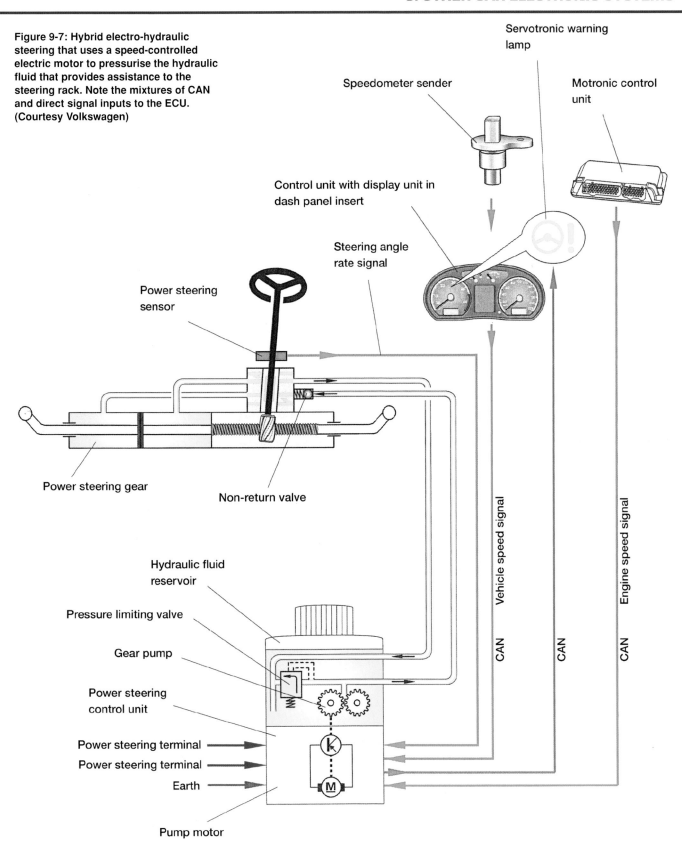

Figure 9-7: Hybrid electro-hydraulic steering that uses a speed-controlled electric motor to pressurise the hydraulic fluid that provides assistance to the steering rack. Note the mixtures of CAN and direct signal inputs to the ECU. (Courtesy Volkswagen)

WORKSHOP PRO CAR ELECTRICAL AND ELECTRONIC SYSTEMS

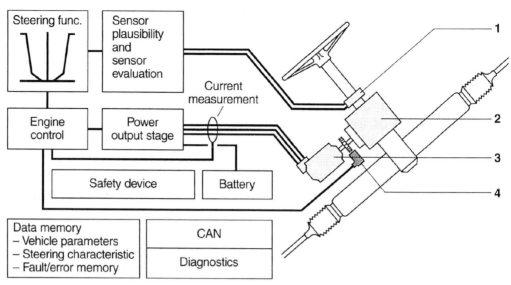

Figure 9-8: Schematic diagram of electric power steering.
(1) torque sensor
(2) reduction gear
(3) electric motor
(4) motor sensor.
(Courtesy Bosch)

ELECTRONICALLY-CONTROLLED POWER STEERING

Many power steering systems use electronic control, either of a system that retains some hydraulic components or by using one that is purely electric.

Hybrid electro-hydraulic steering systems use an electric motor to drive the hydraulic pump, rather than having the pump driven directly by the engine. This approach allows the steering effort to be easily controlled by varying the pump speed. Figure 9-7 (previous page) shows a system of this type.

Electric power-assisted steering completely replaces the hydraulic system. These systems assist driver effort by the use of an electric motor which acts through a reversible gearbox and, in some cases, also an electromagnetic clutch. An ECU determines the degree of assistance that is rendered. Figure 9-8 shows a system of this type.

Electric power steering systems use inputs of steering wheel torque sensors and road speed. The output is to directly drive the electric assist motor (pure electric power steering) or to change the speed of the hydraulic pump (electro-hydraulic steering).

Electric power steering has some significant advantages over any form of conventional hydraulic steering, both for the owner of the car and its manufacturer. The reduction in engine load of an electric power steering system (it can be as low as 4W when the car is being driven in a straight line) means that the fuel economy of a car equipped with electric power steering is very similar to that of a car with no form of power steering. The use of fully electric power steering also allows self-parking and self-driving features to be implemented.

From a manufacturer's perspective, using electric power steering reduces assembly line time, allows the easy software tuning of the steering assistance characteristics

to suit a variety of cars – eg for a sports car or a limousine. (This variation in steering assistance can also be easily modified in cars with electric power steering. For more on this, see the companion volume to this book: *Modifying the Electronics of Modern Classic Cars – the complete guide for your 1990s to 2000s car*.)

Electric power steering can use a number of different mechanical configurations:

Method	Electric assist unit location	Power transmission
Pinion assist	Under the dashboard on the steering column	Motor > worm gear > column shaft > pinion shaft
	On the steering rack input pinion	Motor > gear train > pinion shaft
Rack assist	On the steering rack	Motor > ball screw > rack shaft
	On a second pinion on the steering rack	Motor > planetary geartrain > another shaft pinion > rack shaft

CLIMATE CONTROL

Climate control systems vary in their complexity, depending on the age and original expense of the car. However, in all cases, they attempt to maintain the cabin temperature at the level selected by the driver and/or passengers.

In one system the following inputs and outputs are used:

9. OTHER CAR ELECTRONIC SYSTEMS

In older cars, flaps in the climate control system were often moved by vacuum actuators, but in more modern cars, direct DC motor control is used. The ECU logic may include inhibiting cabin airflow until the coolant is warm enough to provide heat, or the air-conditioning system is cool enough to provide cooling airflow. Some systems also have a road speed input that reduces cabin fan speed as aerodynamic ram effect increases.

Figure 9-9 shows a schematic view of a fairly simple climate control system using vacuum-operated actuators, as might be found in a 1990s luxury car. This type of system does not feature self-diagnostics. Figure 9-10 (overleaf) shows part of the wiring diagram of a more sophisticated climate control system that uses electrically-driven actuators and has links to other car ECUs.

AUTOMATIC TRANSMISSIONS

On many cars, the control of the automatic transmission is integrated into the engine management system. This allows the same input sensors (eg throttle position, intake airflow, engine temperature, etc) to be used in transmission control without the need for duplicate sensors. It also allows the engine operating conditions to be varied as required – eg the ignition timing to be retarded during the gear changes to momentarily drop engine power and so give smoother shifts. On other cars, a dedicated ECU looks after the transmission.

Automatic transmission control is carried out by the actuation of a number of hydraulic valves within the transmission. These control the flow of hydraulic oil which apply and release the internal clutches and bands, causing the gearshifts to take place. The two main inputs used to determine both the internal clamping pressures and when gearshifts occur are throttle position and road speed. Throttle position can be signalled to the ECU by the output of the throttle position sensor, or the ECU may internally model the torque output of the engine (eg by looking at throttle position, airflow, etc) and then use this information. Some older cars had transmissions that, while otherwise electronic, still used a throttle cable that mechanically connected the throttle to the transmission.

Line pressure is also varied within auto transmissions. This pressure controls the clamping forces and has a major influence on when gear changes occur; as engine power output increases, line pressure is increased. Transmissions also have a lock-up clutch in the torque converter, which when engaged stops any slip. This clutch is controlled on the basis of road speed and load, and may also be automatically disengaged when braking.

Transmission fluid control solenoids use two approaches – they're either turned on or off, or they are a variable flow design where the ECU steplessly alters their opening. The solenoids that control the gear change

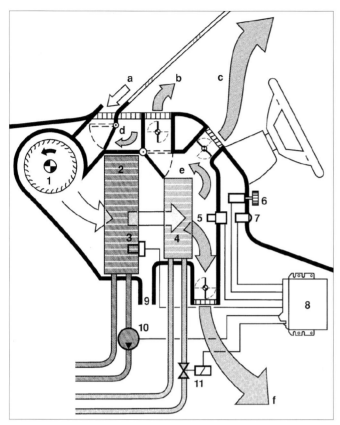

Figure 9-9: A simple climate control system. (1) blower, (2) evaporator, (3) evaporator temperature sensor, (4) heater, (5) air exit temperature sensor, (6) driver temperature adjust, (7) interior temperature sensor, (8) ECU, (9) drain, (10) compressor, (11) heater water valve, (a) fresh air, (b) defrost, (c) ventilation, (d) circulating air, (e) bypass, (f) footwell. (Courtesy Bosch)

Inputs:
- Interior temperature sensor
- Ambient temperature sensor
- Fresh air temperature sensor
- Footwell outlet temperature sensor
- Solar sensor
- Footwell/defrost flap position sensor
- Central flap position sensor
- Temperature mixing flap position sensor
- Fresh air/recirculation flap position sensor
- Refrigerant pressure switch

Outputs:
- Footwell/defrost flap actuator
- Central flap actuator
- Fresh air/recirculation flap actuator
- Blower speed regulator
- Heater flow valve actuator
- Air conditioner compressor clutch relay
- Diagnostic connector

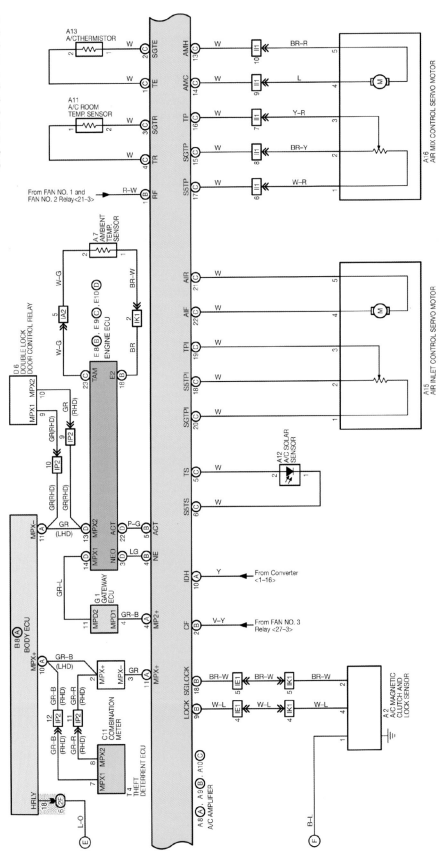

Figure 9-10: Part of the wiring diagram for a Toyota climate control system. The climate control ECU is shown in blue, while the other ECUs with which the climate control communicates are also shown coloured. Notice how the two air flap control actuators use electric motors, with the flap positions monitored by potentiometers. You can also see that ambient temperature is sensed by the engine management ECU that then communicates this information to the climate control ECU. (Courtesy Toyota)

9. OTHER CAR ELECTRONIC SYSTEMS

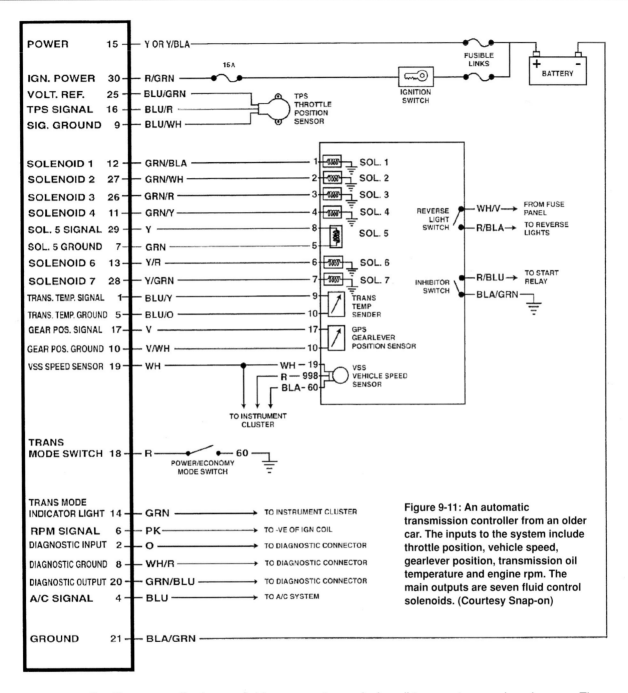

Figure 9-11: An automatic transmission controller from an older car. The inputs to the system include throttle position, vehicle speed, gearlever position, transmission oil temperature and engine rpm. The main outputs are seven fluid control solenoids. (Courtesy Snap-on)

process are generally either on or off, whereas fluid pressure control and torque converter clutch engagement are achieved by steplessly varying the amount of fluid that flows through their respective solenoids. These variations in flow are achieved by varying their duty cycle (ie pulse width modulation).

Figure 9-11 shows a relatively simple electronic transmission control system. The inputs to the system include throttle position, vehicle speed, gearlever position, transmission oil temperature, and engine rpm. There is also a driver-operated power/economy switch. The primary outputs are seven fluid control solenoids, but there are also outputs for a dashboard transmission mode indicator and a diagnostic connector.

Unless it is an externally accessible solenoid or sensor, most faults in automatic transmission control systems rapidly move out of the realm of do-it-yourself fault fixing.

WORKSHOP PRO CAR ELECTRICAL AND ELECTRONIC SYSTEMS

Chapter 10
Fault-finding advanced car systems

- Self-diagnosis
- OBD readers
- OBD codes
- OBD-II readiness indicators
- Fixing systems
- Fault-finding tips
- Seven-step approach

WORKSHOP PRO CAR ELECTRICAL AND ELECTRONIC SYSTEMS

An OBD-II socket. OBD readers plug into this socket and allow you to read and clear fault codes, and see real-time information on the behaviour of the system. The more recent the car and the more expensive the reader, the greater the amount of information that can be gained through this port.

Unlike simple electrical and electronic systems used in older cars, more complex systems have self-diagnosis modes that indicate, via fault codes, what the system 'thinks' is wrong with itself. Depending on the car and system, the fault codes might be able to be ascertained via:
- The flashing of dashboard or ECU warning lights (eg the Check Engine Light)
- A number or message displayed on the dashboard (eg on the instrument panel or entertainment screen)
- A generic plug-in tool (eg Onboard Diagnostics – OBD – covered below)
- A manufacturer-specific specialist service tool (eg the sort of tool you will find in the service workshop of a new car dealership)

Depending on the approach being taken by the system, the code might be a message displayed in English (eg '936 – yaw sensor'), or it might be just the code itself (eg '214', or 'P0037'). As we will see shortly, many OBD codes are universal and so can be easily deciphered, but there are also many codes – probably a majority, in fact – that are manufacturer-specific.

It's also important to note that how you go about fixing a fault is *not* explained by the fault code. Therefore, if you are going to fault-find more complex systems, you will need to have available:
- A means of triggering and reading fault codes for the system you are fault-finding
- Technical information on the system (eg where the components are located, how you can access them, and how you can test them)

It's easy to fall into the trap of assuming that since you can read a fault code, the rest just follows; it doesn't! That said, the presence of a fault code at least gives you a starting point.

SELF-DIAGNOSIS

The earliest cars with self-diagnosis (around the mid-1980s) used flashing lights to communicate fault codes. These are sometimes called 'blink codes.' For example, Nissans of the time used red and green LEDs mounted inside the ECU, which were visible through a little cut-out in the side of the ECU enclosure. One colour LED represented 'tens,' and the other 'units,' so it was just a case of counting how many times each LED flashed to work out the fault code.

(An aside. My fourth car – an expensive car for me, at the time – used the two-LED system in the ECU for its straight-six turbo engine. I bought the car from a dealer and took it home, only to discover the car had a fault code for the knock sensor. There was no Check Engine Light in this car, so you had to manually pull the trim cover away from the ECU and trigger its self-diagnosis function to check the codes. I took the car back to the dealer and complained the knock sensor was causing an issue – and could they fix it under warranty please? After they expressed disbelief that I could tell the knock sensor was faulty – not many owners in those days checked codes – they grumpily agreed to fix it. But they didn't; they just reset the ECU by pulling power, and after a few days the code again appeared. I took the car back to the dealer again. This time they fixed it, all right. They unscrewed the knock sensor from the block, grounded it by running a wire from the knock sensor body to the negative terminal on the battery, then wrapped the knock sensor in foam rubber and cable-tied it in a convenient place. No knock sensor codes now ... and no knock sensor functionality either.)

Other cars of this era flashed the dashboard Check Engine Light to reveal codes. Typically, bridging two terminals in a specific connector activated this function. Depending on where the cars were sold around the world, the connector might have an OBD-style shape (but not necessarily have any OBD functionality) or be a manufacturer-specific connector. If you have a plug that will suit the connector, you can easily make up a device that provides the bridging function. For example, if it's an OBD connector, buy an OBD plug and then wire the correct two pins together. Then, when you insert the plug, the self-diagnosis mode is triggered. The connections that need to

Older cars with OBD-type ports may be able to have fault codes display without the need for a reader. In this case, bridging two terminals in the OBD port (here achieved by using a suitably wired plug) allows codes to be shown on the dashboard.

10. FAULT-FINDING ADVANCED CAR SYSTEMS

be made are shown in the workshop manual for the car, or a web search will probably also find them. If making up such a plug, don't get the connections wrong!

Here is an indicative example of triggering fault codes in this way. It's for a 1999 Hyundai Accent. Its OBD-II socket pins have the following functions:
- Pin 1 – TCM
- Pin 4 – Ground
- Pin 7 – Engine
- Pin 8 – ABS
- Pin 12 – Airbag
- Pin 14 – Vehicle Speed
- Pin 15 – L-wire
- Pin 16 – Battery +

Triggering the fault codes requires these steps:
1. Turn on ignition (do not start car)
2. Ground the L-wire in the data link connector for 2.5 – 7 seconds
3. If no fault is present, '4444' will be flashed on the Malfunction Indicator Light (MIL is another name for Check Engine Light)
4. Each code repeats until the L-wire is again grounded, whereupon the next code appears
5. '3333' indicates the end of code outputs

Most cars of this era used self-diagnosis only for the engine management system, although some more expensive and complex cars also started to use this function for other car systems. However, when you have hundreds (and later, thousands) of fault codes able to be triggered, counting flashing lights soon becomes rather tiresome!

The change in self-diagnosis approach that then occurred was also prompted by government legislation, especially in the USA. In order that cars continued to meet emission standards, the US government mandated that an Onboard Diagnosis (OBD) capability exist in all cars that were being sold, initially in California and then in the whole of the US.

OBD1 was implemented by the California Air Resources Board in 1988 to monitor some emissions components. OBD1 used blink codes through a dashboard light. Subsequently, OBD II was introduced with a standardised connector (officially called a Data Link Connector – DLC), a digital communications port and standardised fault codes (officially called Diagnostic Trouble Codes – DTC).

The SAEJ1979 standard for OBD-II defines a method for requesting various diagnostic data and a list of standard parameters that are available. These parameters are addressed by 'parameter identification numbers' or PIDs. Individual manufacturers often enhance the OBD-II code set with additional proprietary fault codes.

OBD-II has spread around the world and so can be found in the majority of cars of the last few decades. However, note that there is not just one OBD protocol. Different OBD protocols include:
- CAN
- VPW
- PWM
- ISO
- KWP2000

The details of the various protocols need not concern us here, but if you are buying an OBD reading tool, you *must be certain that the tool will work with your specific make and model of car, sold in the market in which you live.* Especially in older cars sold outside of the US, there's a myriad of different pseudo-OBD approaches that were taken by manufacturers – from full OBD compliance to none. Outside of the US, the fact that an OBD-II connector is present in an older car is no guarantee that the car has an OBD-II data stream.

The OBD-II socket is legally required to be located inside the cabin, and must be able to be accessed without tools. (In practical terms, a small screwdriver may be needed to lift an interior trim panel.) If you are trying to locate the socket, start by looking under the dash, behind the ashtray, under the glovebox, under the trim panel beneath the handbrake and in similar places. The 16-pin socket has a characteristic shape (wider at the top than the bottom) and may be protected by a push-on cover.

OBD READERS

Generic OBD readers are very cheap, but be wary of putting down your cash without first investigating the functionality of the tool. In addition to those parameters provided under the SAE standards, as indicated earlier, individual manufacturers use their own extended data read-outs and fault codes. To be most effective, the code reader you use needs to be compatible with as many of the manufacturer-specific codes for your car as possible.

A fault code displayed on the dashboard, after two OBD socket terminals were bridged. Note that different cars use different approaches, so this is not possible in all cars.

WORKSHOP PRO CAR ELECTRICAL AND ELECTRONIC SYSTEMS

This OBD scan tool allows you to display live sensor data, has text-based descriptions of component locations, and supports many generic and enhanced fault codes. (Courtesy OTC)

In addition, you want to be able to access all the different electronic car systems – and there may be dozens of those, from ABS to airbags to transmission control. For example, one particular Snap-on scan tool will read data from each of the following systems in Volkswagen and Audi products:
- Engine management
- Electronic instrument panel
- ABS/EDS/ESP/TCS
- Airbag/pretensioners
- Air-conditioning
- Alarm system interior
- Audio system
- Automatic transmission
- Central door lock system
- Immobiliser
- Steering wheel electronics
- Steering help
- 4WD electronics
- Comfort systems
- Seat adjustment – driver's side
- Seat & mirror adjusting
- Central electronic unit
- Canbus interface
- Additional heater/parking heater
- Electronic level control
- Level control xenon lights
- Tyre pressure monitoring
- Parking help
- Radio
- Navigation systems
- Electronic roof control
- Distance control
- Suspension electronics
- Back spoiler
- Emergency control
- Speech control
- Light control – left
- Light control – right
- Auto light switch

The extent of this list shows the difference in capability between a simple OBD code reader (that can read only generic engine management codes) and the specialised tools used by professional workshops (that can access all car systems and read manufacturer-specific codes).

It's impossible to generalise about which reader will be most suitable for your application. The best approach is to go to an online discussion group that specialises in your make and model of car and ask the people there what they use. In some cases, enthusiasts or private companies have developed new hardware and software to best display operating parameters and fault codes for specific cars and/or manufacturers. Some of these packages can even achieve what is normally a dealer-only processes – changing the action of convenience features, or coding new keys or modules. For example, VCDS for Volkswagen/Audi cars is, as the seller says, "a software package for Windows that emulates the functions of the dealers' very expensive proprietary scan tools." When I owned a Volkswagen product (a Skoda), I used VCDS. I have a 2000 model hybrid Honda Insight that I've extensively modified. One Insight enthusiast, Peter Perkins, has developed hardware and software that allows the reading of all fault codes on this car, even those Honda-specific ones generated by the hybrid control system.

Typically, to be most useful, the OBD reader needs to be 'less generic' and 'more specific' to your car.

Generic OBD displays are available as colour dashboard displays, including head-up displays that show the readings reflected on the inside of the windscreen. However, such displays are less useful as diagnostic tools because they typically display only a quite limited number

A professional level scan tool such as this one will allow access to nearly all car systems, and read manufacturer-specific as well as generic OBD codes. This unit also incorporates a multimeter and oscilloscope. It's also out of financial reach for most people working on their cars at home! (Courtesy Snap-on)

10. FAULT-FINDING ADVANCED CAR SYSTEMS

of parameters. While not as flashy as some of the readers that are available, ScanGauge has a good reputation as a general-purpose OBD reader and display, and can clear fault codes. I have used a ScanGauge and it worked just as advertised.

OBD readers simply plug into the OBD socket – no other installation is needed.

OBD CODES

OBD codes use a specific format. These are shown in the table below.

First digit	Systems	B=Body C=Chassis P=Powertrain U=Network
First and second digits	Code type	Generic (SAE): P0 P2 P34-P39 B0 B3 C0 C3 U0 U3 Manufacturer Specific: P1 P30-P33 B1 B2 C1 C2 U1 U2
Third digit	Sub-system	1= Fuel and air metering 2= Fuel and air metering 3= Ignition system or engine misfire 4= Auxiliary emission controls 5= Vehicle speed control and idle controls 6= Computer output circuits 7= Transmission controls 8= Transmission controls

For example, a 'P0129' code is a powertrain, fuel and air metering code – in fact, it indicates that a barometric pressure sensor is too low in output.

OBD-II READINESS INDICATORS

Readiness indicators show the results of tests that the system performs on itself. Some of these are continuous (they're running all the time) while others are performed only under certain conditions.

Continuous monitors comprise:
- Misfire
- Fuel system
- Comprehensive Components Monitor (CCM)

The comprehensive component monitor (CCM) is a diagnostic program run by the engine management control unit. It checks for open circuits, shorts to ground, shorts to power and the rationality of the signals coming from the sensor circuits. The decision about the rationality of the signal is dependent on operating conditions. For example, if the engine rpm and throttle angle are both low, but the air mass-flow is very high, the air mass reading is not rational for that rpm and throttle angle. In this situation, a fault code for implausible air mass will be stored.

Non-continuous monitors are:
- EGR system
- O2 sensors
- Catalyst
- Evaporative system
- O2 sensor heater
- Secondary air
- Heated catalyst
- Air-conditioning system

OBD systems indicate if these tests have been completed via 'Ready' or 'Complete' tags. For the OBD monitor system to become complete, the vehicle should be driven under a variety of normal operating conditions. These operating conditions may include a mix of highway driving, city driving and at least one overnight ignition-off period. Readiness indicators are designed to allow inspection stations to be able to quickly pass (or not pass) cars that need to meet emissions standards while in use.

As the main text suggests, OBD refers specifically to Onboard Diagnostics. An OBD code reader would therefore, you'd think, only be able to read OBD codes. However, although only emission-related codes and data are required by legislation to be transmitted through the OBD port, most manufacturers have made this connector the conduit through which all systems are diagnosed. Therefore, I am using the term 'OBD reader' as a generic term for an instrument that can read data from this port. Sometimes, these readers are also called 'scan tools.' In addition to reading fault codes on a scan tool, live data from inputs and outputs can usually be displayed.

FREEZE FRAMES

When an emissions-related fault occurs, the OBD-II system not only sets a code but also records a snapshot of the vehicle operating parameters to help in identifying the problem. These values are known as Freeze Frame Data. The data may include engine rpm, vehicle speed, airflow, engine load, fuel pressure, fuel trim value, engine coolant temperature, ignition timing advance, or closed loop status.

FIXING SYSTEMS

So how do you go about fixing a fault in a complex car system – whether it's engine management, climate control, automatic transmission control or stability control?

Firstly, what does the fault code show? Let's take an example. You have an indication of an engine problem (eg Check Engine Light on) and you do an OBD scan that reveals a P0118 code. A quick web search under this fault code shows that indicates a problem with the Engine Coolant Temperature (ECT) sensor, and specifically that the output is high.

Using your OBD reader, check that the temperature of the coolant matches that shown on the reader. That is, if the coolant is actually at 85°C (185°F) then the reading that the ECU is seeing (as shown by the diagnostic tool) should be within 5°C (9°F) of this. (The actual coolant temperature can be checked with an external thermometer.) If in fact the displayed and actual temperatures are in fairly close agreement, then the wiring to the sensor is probably fine – an open or short circuit will result in a reading that is massively in error. If on the other hand, the sensor's monitored temperature is quite different to the value it should show, check the wiring and the plug on the sensor for corrosion.

The next step is to remove the sensor and check that its resistance varies with temperature appropriately, ie that the sensor changes in resistance with temperature as shown in the workshop manual. If it doesn't, replace the sensor.

Because there are thousands of OBD fault codes, the web is your fastest method of finding out what the code represents, and may then give you some instructions on how to solve the problem. www.troublecodes.net is a good site, and it gives the following information about the P0118 code we're using here as an example:

Common Causes of P0118:
- Bad connection at the sensor
- Short in the voltage feed between the sensor and ECU
- Defective ECU (not likely)
- Defective ECT sensor (shorted internally)

What happen when P0118 sets:
- Cooling fans will be commanded ON
- Engine coolant temperature gauge is inoperative
- AC compressor will be commanded OFF

Conditions setting P0118:
- ECU detects that the ECT is colder than -38°F (-39°C) for more than five seconds.
- Ignition is ON or engine is running for more than 10 seconds.
- Engine run time is less than 10 seconds when IAT sensor 1 is warmer than -7°C (19°F).
- DTC P0118 runs continuously within the enabling conditions.

The car-specific manuals available from major makers of OBD readers and scan tools also give lots of details about the data able to be read from specific vehicles. This material can include not only the manufacturer-specific fault codes, but also what the various monitored parameters mean. The information below is an example of the material available in the 'Mercedes-Benz Vehicle Communication Software Manual' from Snap-on Tools. This manual, and similar ones covering many other cars, are available as free downloads from Snap-on.

INTAKE AIR TEMPERATURE
Range: -60 to 65°C or -76 to 150°F
Used on DC12, DM, DM2, EDS, ERE/EVE/ASF (IFI Diesel)), EZ, HFM, LH, ME10, ME20, ME27, ME28, and SIM4 systems. This parameter displays the temperature of air coming into intake manifold in °C or °F. Reading is based on the input signal of the intake air temperature (IAT) sensor. On Diesel system ERE/EVE/ASF (IFI Diesel)), this parameter is used for fuel metering control, which limits smoke emissions, for controlling EGR, and for intake pressure control. The measurement units can be changed from degrees Celsius (°C) to degrees Fahrenheit (°F). The preset measurement is °C.

INTAKE MANIFOLD SW.-OVER VALVE
Range: ON/OFF
Used on DM2 and HFM and ME20 systems. These parameters display the state of the resonance flap used in the air induction system. When the display reads OFF, the flap is closed with the engine running at low speeds. When the display reads ON, the flap is open with the engine running at high speeds. The pneumatically controlled resonance flap is located on intake manifold, and effectively creates two different intake manifold lengths. The resonance flap is connected to the intake manifold switchover valve, which is controlled by the ECU. At low engine speeds, with the resonance flap closed, air is directed into the longer intake runners. This increases low-end torque by using the ram air effect. At high engine speeds, with the resonance flap open, intake air is fed into the short intake runners. This increases the volume of air to meet the higher demands of the engine.

10. FAULT-FINDING ADVANCED CAR SYSTEMS

Other downloadable manuals from Snap-on describe the location of individual cars' diagnostic connectors, the ways to trigger blink or OBD codes, and the step-by-step of testing different systems. These manuals are a fantastic resource, especially if you have an older car without full OBD capability.

When searching on the web for information that may help you, don't forget model-specific enthusiast discussion forums. While I must say that personally these discussion forums sometimes drive me nuts, they're also a useful source of information on common faults. For example, earlier I mentioned my Honda Insight; it's a car renowned for having odd electrical problems caused through the breakage of the ground straps between the engine and bodywork in the engine bay. Unless you were alerted to this, that's not at all the first place you'd go looking for problems! Similar 'inside' knowledge is available at lots of other car-specific groups.

FAULT-FINDING TIPS

Don't become transfixed by the fault code. The fact that the system indicates a fault with a particular sensor or controller, for example, does not always mean that is the part actually at fault.

A specific example. My Mercedes E500 uses two 12V batteries, with current flow between them controlled by an electronic gateway. A fault diagnosis revealed that this gateway was suspect – and it was an expensive proposition to buy another. However, replacing one of the 12V batteries fixed the problem! Rather than the gateway being at fault, it was in fact the 12V battery that was getting tired.

In many cars, a cruise control system fault is most commonly traced to a faulty brakelight or brakelight switch. Rather than dive into the cruise control system, it's best to make some basic checks of the car first.

If a fault code indicates that an input sensor is defective, the ECU (especially in older cars) usually has no way of knowing whether it is in fact the sensor that is faulty, or if it is the wiring that connects to that sensor. Rather like the idea that some people have that, if a light in their house is faulty, it must be the wall switch (after all, the switch no longer operates the light, does it?), a fault code indicating a specific sensor failure is often best thought of as a *fault in that part of the system*. Therefore, firstly look at that system as a whole, rather than homing straight in on the alleged faulty sensor.

Some sensors age and fail. Vane airflow meters, throttle position sensors and oxygen sensors degrade over time. These are quite likely to need replacement.

On the other hand, unless they are physically falling apart, knock sensors and intake air temperature sensors are much less likely to fail. Especially in older cars, wiring, plugs, sockets, and relays are all candidates for problems. Moving parts and actuators in systems like climate control are notorious for failure, whereas for example the outside temperature sensor is not. Concentrate on the most likely problem areas before randomly changing parts to see if the problem is fixed.

That said, in some cases, trial replacement of parts is the only way. I once had a complex car (twin turbo, all-wheel drive) that developed an engine miss, but only after being driven continuously at highway speed for 2-3 hours. There were no fault codes – and no obvious solution. The problem was serendipitously fixed when I installed a plug-in aftermarket ECU – almost certainly, the original ECU had an intermittent problem that developed only after running for that length of time (or maybe the aftermarket ECU used a lower coil dwell time, and it was actually a coil causing the problem? No matter – changing the ECU fixed it!)

STEP-BY-STEP

When confronted by faults in complex car systems, I suggest you take the steps shown in the following list. In presenting the list I am conscious that some people will consider these steps obvious, but over the years I've seen many people spend a lot of money replacing parts that were not actually faulty. Had they followed these steps, that would not have happened!

Step 1: Scan the fault code(s).

Step 2: If the fault code does not have any related symptoms (eg poor driveability), clear the fault code and see if the code reappears. If it does not reappear, put it down to a momentary glitch. If it does reappear, move to the next step.

Step 3: Research on the web, searching for information on this fault in your particular model of car. Post to relevant discussion groups, citing the fault code and the symptoms (if any) and asking what is most commonly the cause of the problem.

Step 4: Carefully inspect the part of the system that the fault codes and web research indicates is suspect. Consider also other systems that may be linked and may be causing the problem.

Step 5: Make measurements eg wiring continuity, checking that voltages and resistances are as per the specs in the workshop manual, looking at scope traces for position sensors.

Step 6: Replace parts or fix wiring.

Step 7: Clear the fault code and ensure it does not reappear.

WORKSHOP PRO | CAR ELECTRICAL AND ELECTRONIC SYSTEMS

AN OLD AND COMPLEX CAR

It's impossible to cover every scenario where something electrical or electronic in a car needs to be fixed, but here's an example of the approach that I take. The fixes were performed on the electric seats of an old E23 BMW 735i (dating right back to 1985!). The seats did not have a self-diagnosis function and I didn't have a workshop manual. Note that with these electrical/electronic/mechanical systems in older cars, it's usually a mechanical and/or wiring problem that's causing the issue. Again, as is typical, in the BMW the engine was running fine but the body electrical systems were showing their age.

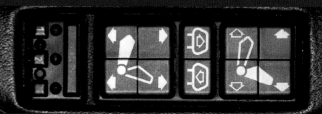

The seat functions are all controlled by this complex switchgear, which is duplicated on the passenger side of the car. As can be seen here, the driver's seat also has three position memories. Incidentally, the rear seat is also electric, with an individual reclining function for each side! But in this car, the back seat was working just fine.

The electric seats

The front seats in the BMW are among the most complex that you'll find in any older car. They have electric adjustment for front/back travel, front of the seat up/down, rear of the seat up/down, head restraint up/down and backrest rake forwards/backwards. However, they don't have electric lumbar adjust and they don't have airbags. (If the seats that you are working on have airbags you must read the factory workshop manual to ascertain the safe procedure for working on the seats.)

The driver's seat had three problems.
- The button which moved the seat rearwards didn't work. However, the seat could be moved backwards with one of the memory keys.
- The front of the seat couldn't be raised.
- The head restraint wouldn't move up or down, although in this case the motor could be heard whirring uselessly whenever the right buttons were pressed.

10. FAULT-FINDING ADVANCED CAR SYSTEMS

The first step was to remove the seat from the car so that access to all the parts could be gained. The seat was electrically moved forward and then the two rear floor-mounting bolts undone.

Using a home-built, heavy-duty, over-current protected, 12V power supply, power was applied to pairs of terminals connecting to the thick wires until the right connections were found. The seat was then powered backwards until the front mounting bolts could be accessed. These were removed and then the seat moved forward until it sat in the middle of its tracks, making it easier to get out of the car.

But how was access going to be gained to the front mounting bolts? Pressing the adjustment button didn't cause the seat to move backwards, and by this stage the memory button had stopped allowing that action as well! The answer was to manually apply power to the seat to activate the motor. All the connecting plugs were undone and those plugs containing the heaviest cables inspected. (There will be wiring for seat position transducers and things like that in the loom, but the motors will be powered by noticeably heavier cables.)

This is what the BMW seat looks like underneath. Four electric motors can be seen, plus there's a fifth inside the backrest. Each electric motor connects to a sheathed, flexible drive cable that in turn connects to a reduction gearbox. As I later discovered, inside each gearbox is a worm that drives a plastic gearwheel, which in turn drives a pinion operating on a rack. At this stage, though, a simple test could be made of each motor by connecting power to its wiring plug and making sure that the function worked as it should. Every function but the head restraint up/down worked, so the problems other than the head restraint showed that they must be in the switches and/or plug connections, not the motors or associated drive systems. But how to fix the head restraint up/down?

WORKSHOP PRO CAR ELECTRICAL AND ELECTRONIC SYSTEMS

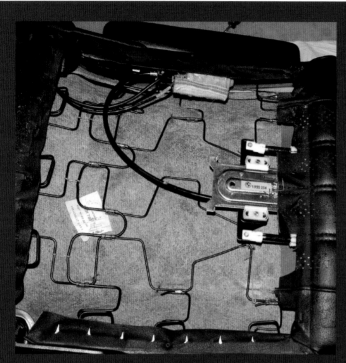

The rear trim panel of the seat came off by simply undoing of four screws. As with the other seat motors, the mechanism consisted of a brush-type DC motor driving a flexible cable that went to the adjust mechanism. The motor worked fine with power connected, but the head restraint didn't move. Feeling the outside of the drive cable sheath indicated that the drive cable inside was turning, so the problem must lie in the mechanism closest to the head restraint itself.

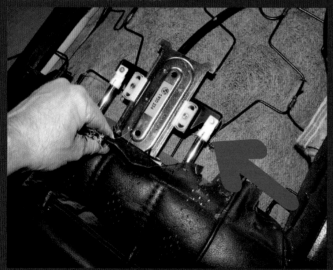

The adjustment mechanism was held in place with one screw, which was accessible with the leather upholstery disengaged from small metal spikes that held it in place. The legs of the head restraint clipped into plastic cups on the mechanism (one is arrowed here) and these were able to be popped out with the careful use of a screwdriver.

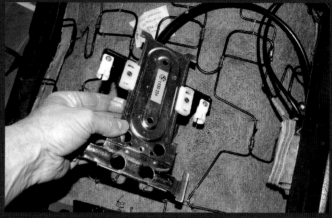

The whole upper part of the adjustment mechanism was then able to be lifted out of the seatback and placed next to the seat. Note that the electric motor stayed in place – it didn't need to be removed as well.

The gearbox was held together with four screws, each with an oddly-shaped socket head for which I don't have a tool. However, knowing that I could always find replacement small bolts, I used a pair of Vicegrips to undo them – that is, it didn't matter if they got a bit mutilated in the process of disassembly.

To see what was going on inside the unit, it needed to be pulled apart. It was obviously never designed to be repairable, and so the first disassembly step involved drilling out the rivets which held the plastic sliders in place on their track. With these out, the action of the pinion (a small gear) on the rack (a toothed metal strip) could be assessed. Neither looked particularly worn and applying power to the motor showed that in fact the pinion wasn't turning. So that meant that the problem was inside the gearbox itself.

10. FAULT-FINDING ADVANCED CAR SYSTEMS

Inside the gearbox the worm drive and its associated plastic gear could be seen. Initially I thought that the plastic cog must have stripped, but inspection showed that this wasn't the case. So why wasn't drive getting out of the gearbox? Again I applied power to the motor and watched what happened. What I found was although the cable could be heard rotating inside its sheath, that drive wasn't getting to the worm. Pulling the worm gear out and inspecting the square-section drive cable showed that the end of the cable was a little worn and it was slipping back out of the drive hole of the worm. (The slippage was occurring inside the area marked by the arrow.)

The mechanism could then be reassembled. New screws were used to replace the Vicegripped ones, while the drilled-out rivets were also replaced with new screws and nuts (arrowed). The gearbox was re-greased before assembly and a smear of grease was placed on the tracks that the nylon sleeves run on. Back in the seat, the mechanism was checked by applying power – and worked fine.

Since all the motors had now been proved to be in working order, fixing the electric rearwards travel and front up/down motion could only be achieved with the seat back in the car – it looked as if it had to be a wiring loom or switchgear problem. But while the seat was out, it made sense to wipe over all the tracks and exposed cogs and re-grease them.

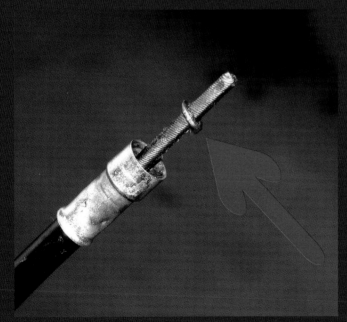

The fix was dead-easy – simply pull the drive cable out of the sheath a little, crimp a spring steel washer on it (backed by a plain washer that here is out of sight – it's fallen back into the mouth of the sheath) and then push the drive cable back down in its sleeve. With the crimped washer preventing the worn part of the cable from sliding back out of the square drive recess in the worm, drive was restored to the gearbox.

Under the driver's seat is a control module containing both relays and the seat memory facility. Close inspection of the plugs and sockets on both the unit and the associated loom showed that some corrosion had occurred. (Perhaps at some stage a drink had been spilled on it.) The corrosion showed itself as a green deposit on the pins and some tedious but careful scraping with a small flat-bladed screwdriver removed it. Once that was done, the associated plug was inserted and pulled out 20-30 times to scrape off the deposit inside the pins of the plug, which were otherwise impossible to access to clean. This done, all seat functionality was restored.

WORKSHOP PRO CAR ELECTRICAL AND ELECTRONIC SYSTEMS

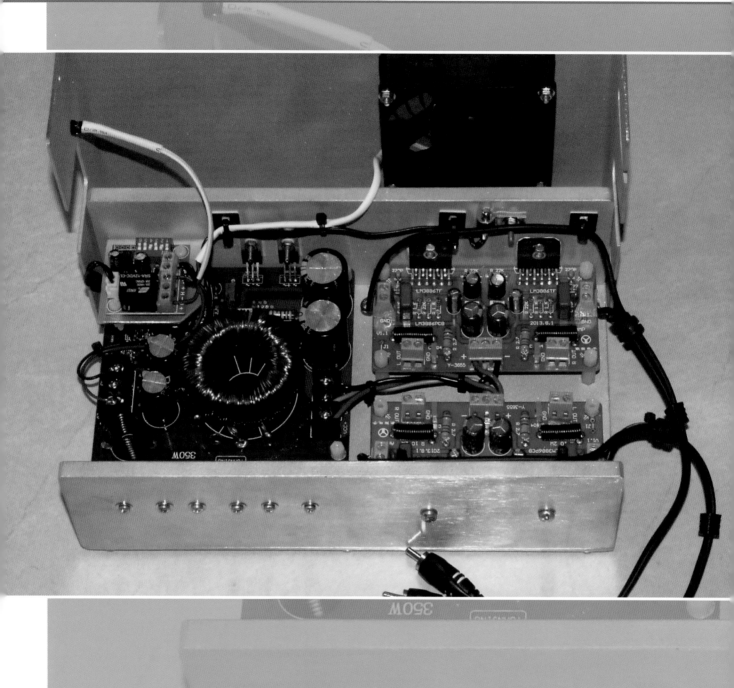

11. ELECTRONIC BUILDING BLOCKS

Chapter 11

Electronic building blocks

- Timer modules
- Temperature controller and display
- Voltage switch
- Smart battery monitoring LED
- Voltage booster
- High current flasher
- Variable voltage power module
- Adjustable voltage regulator
- Four-channel amplifier
- Multi-purpose module
- Stemsel

WORKSHOP PRO: CAR ELECTRICAL AND ELECTRONIC SYSTEMS

One of the great changes in car electronics – and hobbyist electronics in general for that matter – is the availability now of high quality, pre-built electronic modules. They're especially good in car modification, because they let you achieve outcomes that previously would have been much more expensive and difficult. All the modules can be bought online, and even if the exact module is no longer available by the time you read this, very similar modules will be.

If you have not worked with electronic modules before, keep in mind that:
- As supplied, these modules comprise just a built printed circuit board. In use, the controller must be mounted in a box. Be careful when working with the bare module that the underside of the board cannot contact anything metallic – this will create short circuits and destroy the module.
- Most modules are not protected against reverse polarity connection – if you get the positive and negative terminals around the wrong way, you'll probably blow-up the module. Those modules with transistor outputs are usually not protected against short-circuiting the output (the eLabtronics Multi-Purpose Module, to be covered later, is a notable exception to this). Modules that use relay outputs are more forgiving!
- The modules should always be fed power via a fused supply.

Most of these modules are incredibly cheap, so in many cases, it's worth buying two or three of them at a time.

GENERAL PURPOSE TIMER MODULE

Need to operate something for a period at the press of a button? It doesn't get much cheaper or easier than this! Available is this general purpose 12V timer. It's adjustable for periods from 1 second to 3 minutes, comes with an onboard relay capable of running devices drawing up to 10A, is fully built and easy to wire into place. Search online for 'Adjustable Delay Timer Module 12V.'

Measuring only 56 x 26 x 23mm (L x W x H), the tiny module has an LED, relay and multi-turn adjustable pot. There are two terminal strips, one at each end. At one end there are the connections for:
- 12V and ground (don't get them around the wrong way)
- Trigger input

At the other end there are the relay connections for:
- Common (x2)
- Normally open
- Normally closed

The connections are easy. Let's say that you want to run a windscreen washer spray for 2 seconds with just one push of the button. In that case connect:
- 12V terminal to ignition-switched 12V
- Ground terminal to the chassis
- The two trigger wires to a normally open pushbutton (doesn't matter which way around these wires go)
- One side of the pump to ignition-switched 12V
- The other side of the pump to the relay's normally open terminal
- The common of the relay terminal to the chassis

With 12V, ground and the pushbutton connected, you can easily set up the timer. When the pushbutton is pressed, the relay should click and the LED come on. When the timed period has finished, the LED goes out and the relay again clicks (as it switches off). The multi-turn adjustable pot can be turned with a small blade screwdriver to change the timed period. As viewed from above, turn the pot clockwise to make the timed period shorter (and of course, anti-clockwise to make the timed period longer).

You can mount the module in a box, but cheaper and easier will be to just wrap it in electrical tape or enclose it in large diameter heat-shrink. Do this only when you have the timed period set correctly!

PULSING TIMER MODULE

Here's a great electronic module that can be used to pulse an output. It is widely adjustable, with both the 'on' and 'off' times able to be set separately. It uses a relay output that is able to drive high current loads (up to 10A). Finally, it's incredibly cheap. Do an online search under '12V DC Circulate Time Delay Relay module.'

The fully constructed module is about 56 x 30mm. At one end of the board it has inputs for power and ground, and the other end has three relay connections – one for common, one for normally open and the other for normally closed. There are two pots mounted on the board – one controls the 'off' time and the other the 'on' time. A red LED glows whenever power is applied, and a green LED turns on when the relay is activated. There is also a configurable link on the board. In a very tricky move, placing the link on the board feeds 12V to the 'common' terminal of the relay. This makes wiring much simpler in many applications, because the load can be connected between the other

A very cheap general-purpose timer with a relay output. This module is adjustable for periods from 1 second to 3 minutes.

11. ELECTRONIC BUILDING BLOCKS

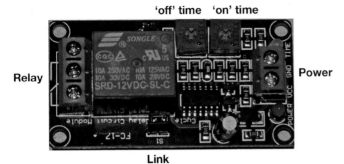

This electronic module can be used to pulse an output. It is widely adjustable, with both the 'on' and 'off' times able to be set separately. It uses a relay output that is able to drive high current loads (up to 10A).

relay terminal and ground. The board is well made and a quality relay is used.

The easiest way to see how the module works is to connect power and ground to the appropriate terminals. Rotate both pots fully anti-clockwise. Turn on power and the red LED will immediately light. With the pots set as described, the green LED will flash (and the relay click), with the 'on' and 'off' times both being 1 second. This is the shortest output time available.

Rotate the 'off' pot a fraction clockwise. The output will still be activated for one second but this might occur now only every 5 seconds. And if for example you wanted the output to be activated at 5 second intervals for the longer period of 10 seconds, you'd turn the 'on' pot a little more clockwise.

You can see that both the frequency and duty cycle can be adjusted in this way.

On the sample module, the 'off' time was adjustable from 1-120 seconds, and the 'on' time from 1-40 seconds. When the pots are adjusted correctly for the application, a dab of nail varnish can be used to hold them in the chosen positions.

Wherever you want to pulse an output, think of this handy module!

TEMPERATURE CONTROLLER AND DISPLAY

Do you want a dashboard display of the temperature of your engine's coolant, oil or transmission temperatures? And you'd also like to be able to sound an alarm or illuminate a warning light if the temperature exceeds a pre-set threshold? This module can do all of that – and cheaply too. It is available in both Celsius and Fahrenheit versions – I'll use the Celsius version as the example here.

The module is 78 x 71 x 29 mm (L x W x H) and uses a display window that requires a cut-out 70 x 28mm. It has a mass of 110 grams. It uses an LED display that shows temps up to 100°C to one decimal place (eg 35.6), and above 100°C in single units (eg 105). The update rate is fast (about three times a second) and the sensor is very responsive to changes in temperature.

In addition to the numerical display, there are two individual LEDs. One shows when the set-point has been exceeded. (The set-point is the temp at which you've set the device to activate its output.) This LED has two modes – steadily on when the relay is activated, and flashing when the set-point has been passed but the module is running an inbuilt delay before turning on the output. (You can vary this delay time – more on this in a moment.) The other single LED shows that the display is indicating the set-point temperature. On the face of the instrument are four push buttons – up/down arrows, Set and Reset. Wiring connections are by means of screw terminals on the rear of the module.

The simplest use of the instrument is to display just temperature. This requires only four wiring connections and no menu configuration. Power (12V nominal) connects to Pins 3 and 4 – ground to pin 3 and positive to pin 4. The NTC (Negative Temperature Coefficient) sensor that is provided connects to pins 7 and 8 – it doesn't matter which wire goes to which terminal. With these connections made, the display should come alive and show the temperature at the sensor. (Note: the thermistor wiring can be extended as required. If you were to use it to sense oil temperature or similar, you would need to make an appropriate screw-in fitting and glue the sensor within it, using high temperature epoxy.)

The module is fitted with a 5A relay. This means it can be connected directly to low voltage buzzers, fans and warning lights. To get a feel for how the control system works, it's a good idea to play with it before installation. Let's take a look at how it can be set up.

Pressing the Set button briefly changes the display to show the set-point temperature. This setting can be altered by pressing the up and down keys. When done, press the

This quality temperature display and controller is ideal for controlling fans and warning lights, with the bonus that it also shows the temperature being monitored. It's available in both Celsius and Fahrenheit versions.

WORKSHOP PRO CAR ELECTRICAL AND ELECTRONIC SYSTEMS

Set key again or simply wait a few seconds and the display reverts to the current temperature.

Pressing the Set button for three seconds brings up a second menu. Different parameters can be selected by pressing the up/down keys. To change the selected parameter, press the Set key a second time then make the adjustments with the up/down keys. Whatever setting is selected is retained in memory, even if power is lost.

The available parameters are:

HC – this menu configures the module to either turn on its relay when the temperature exceeds the set-point ('C' mode), or turns on the relay when the temp falls below the set-point ('H' mode).

d – this sets the difference in temp between switch-on and switch off. (This is sometimes called the hysteresis.) By using the up/down keys, you can set this anywhere from 1°C to 15°C. This is a very powerful control that can make a huge difference to how the system functions.

L5 – this is the minimum temperature the set-point can be configured. Normally, this would not need to be altered from its minus 50°C default.

H5 – this is the maximum temperature the set-point can be configured. Normally, this would not need to be altered from its 110°C default.

CA – this function allows you to correct the temperature display by adding or subtracting 1 degree units from the displayed reading.

P7 – this function is used when in C mode you don't want the output cycling on and off at short intervals. The setting can be anything from 0-10 minutes. It example, if it is set to 1 minute, after the relay has activated once, it will not activate again until a minute has passed – even if the temp set-point has been tripped. In most uses you would set this to zero.

VOLTAGE SWITCH

This digital voltage switch can monitor the output of any existing car sensor that outputs a voltage in the range of 0.5-5V. That includes throttle position sensors, most airflow meters, most oxygen sensors, fuel level senders, intake air and coolant temperature senders, oil pressure senders and others. No more trying to fit a sensor that was never designed to be there – you just make use of the existing factory sensor!

Whatever variable is already being sensed – whether that's engine load, engine temperature, oil pressure, fuel level, etc – can now be used to additionally switch something on and off. So you can trigger radiator cooling fans (using the coolant temp sensor), turn on intercooler water sprays at high loads (using the intake air temp sensor), switch on intercooler fans at high engine loads (using a throttle position sensor), sound a low oil-pressure alarm if the oil pressure drops (using the factory oil-pressure sensor), and so on. You don't need to disconnect the ECU or dash – just tap into the signal with the voltage switch!

LOADING-DOWN SENSORS?
The maximum current drawn from the sensor by the module is only 50µA– this is very, very small. I'd expect that with such a low current draw, you could monitor any voltage-outputting sensor on the car without changing its output or affecting its operation.

Unfortunately, the instructions that come with the module that I bought are not very clear, but I'll address that problem here.

The module is fully built, and is sized at 67 x 44 x 20mm. It has four onboard pushbuttons:
- Set
- Sw1
- (+)
- (-)

It has a three-digit LED display that shows the monitored voltage, and two onboard LEDs: one red (power on) and one blue (relay tripped). The relay is a single pole, double throw unit rated at 10A at 30V DC. The electronics on the board are surface mounted.

At one end of the board is a four-terminal screw-type connector. The connections are:

DC+	positive supply (battery 12V)
DC-	negative supply (chassis or ground)
V+	the positive of the monitored voltage (eg to sensor output signal)
V-	the negative of the monitored voltage (chassis or ground)

This adjustable voltage switch can be used to monitor the output of any sensor in a car that outputs between 0-5-5V. The onboard relay can be used to switch loads at preset voltages.

At the other end is a three terminal strip for the relay – C (common), NO (normally open) and NC (normally closed). When the relay closes, the C and NO terminals are connected.

The functionality of the module is programmable. There are five different functions available – F1 through to F5. Also programmable are two voltage levels – P1 and P2.

The functions are shown below:

Function	Description
F1	Voltage display only. No relay operation.
F2	Relay on when voltage below P1. Does not go off until voltage rises above P2.
F3	Relay on when voltage above P2. Does not go off until voltage falls below P1.
F4	Relay on when voltage between P1 and P2. Off at other voltages.
F5	Relay off when voltage between P1 and P2. On at other voltages.

So what are the implications of these functions? Here are some examples:
- If you wanted an intercooler water spray to come on at high throttle angles, you'd select Function 3 and use the output of the throttle position sensor.
- If you wanted a light to come on when the engine was either cold or hot (ie is outside of normal operating range) you'd use the output of the coolant temp sensor and Function 5.

To select the Function, do the following:
1. Long press Set until P-0 shows
2. Short press Set
3. Use + and – buttons to select correct Function (F-1 to F-5)
4. Long press Set until voltage display resumes

To set P1 (lower voltage):
1. Long press Set until P-0 shows
2. Press Sw1 until P-1 shows
3. Short press Set
4. Short press Sw1 until the correct digit flashes
5. Use (+) and (-) keys to change displayed digit
6. Repeat steps 4 and 5 until the correct voltage has been set
7. Long press Set until voltage display resumes

To set P2 (higher voltage):
1. Long press Set until P-0 shows
2. Press Sw1 multiple times until P-2 shows
3. Short press Set
4. Short press Sw1 until the correct digit flashes
5. Use (+) and (-) keys to change displayed digit

Two other functions
Calibration
P3 (accessed in the same way as P1 and P2) allows the voltage display to be calibrated. Double presses of the (+) and (-) keys are needed to change tenths of a volt.
Display on/off
Short press of Set switches off LED. Short press of Set turns it back on.

6. Repeat steps 4 and 5 until the correct voltage has been set
7. Long press Set until voltage display resumes

If you are operating high currents (eg a radiator fan), you'd need to use an additional relay, as the current draw of the fan is too high for the onboard relay. However, the onboard relay is fine for powering warning lights, a buzzer, intercooler water spray pump, etc.

The minimum hysteresis (the difference between switch on and switch off voltages) is 0.2V – that's the case even with P1 and P2 set to the same number. This hysteresis prevents the relay from chattering. Of course, by setting P1 and P2 to voltages more than 0.2V apart, you can set the hysteresis to be much greater than 0.2V.

SMART BATTERY MONITORING LED

Here's a brilliant battery monitor. It is made by Gammatronix in the UK, and is available via eBay. To find it, search for '6v, 12v, 24v Programmable LED Battery level voltage monitor meter indicator.'

At first appearance, it looks just like a 10mm LED mounted in a bezel. But then when you look closer, you'll see a tiny pushbutton – and if you pull the assembly from the bezel, you'll also see a programmable chip and a few other components. In fact, what we have here is a 6, 12 and 24V battery monitoring LED that is user-programmable to run one of six different in-built maps. The single LED can show green, red, yellow (a yellow that actually looks more orange) and white (off). The LED can also flash at different rates. The voltage is monitored as a rolling average over

At a glance, the battery monitor just appears to be a 10mm white LED. However, its functionality is much greater than it first appears. The arrow points to the pushbutton that allows the LED to be programmed as a smart battery monitor.

2 seconds, is claimed to be accurate to 1 per cent, and will operate over the range of 3.8-30V. Wiring is as simple as you can get – red to positive and black to negative.

So what are the different in-built battery monitoring maps that are available? There is one for pretty well every use you can think of.

Here are the different maps:

Map 1: Battery Voltage Monitor

This is the factory default voltage indicator mode. Low distraction, minimum of colour changes in normal operation, suitable for vehicle use, such as motorcycles, cars, boats, campers, etc. By flashing and using different colours, it shows eight different voltage ranges from 10.5 to 15+ volts.

Map 2: Vehicle Charge Indicator

This map illuminates the LED green when under charging conditions, ie when the vehicle alternator is working. Yellow and red will show if the battery is discharging. This mode monitors seven different voltage ranges from 10.5 to 15+ volts.

Map 3: Vehicle Monitor, includes fake alarm

This is great for motorcycles, and vehicles stored long-term. When riding/driving and charging, the LED is steady green. 30 seconds after charging stops (ie the vehicle is parked), the unit will enter low current mode to show battery status while the vehicle is in storage. The LED will blink green, yellow or red to show stored state battery condition. An added benefit is that LED blinking looks like a vehicle alarm. This mode has only a very small current draw (0.5mA) from battery.

Map 4: Charge Indicator (Stealth Mode)

This is similar to mode (2) but the LED is not illuminated under normal charging conditions. That is, the LED is blank in normal operation. Yellow and red illuminations signal charging faults or discharging battery.

Map 5: High Res 10 step voltage monitor

This mode is a high-resolution mode where maximum resolution is important and colour changes and flashing are not distracting. This mode monitors 10 different voltage ranges from 10.5 to 15+ volts.

Map 6: Minimal Monitor

This mode uses a simple low current (less than 0.5mA), three colour battery status monitoring. A short flash every 2 seconds indicates current state, with 5 different voltage ranges from 10.5 to 15+ volts.

So is it all good news? Well, I initially found the instructions rather hard to understand - especially in the area of mode set-up. Let's take a look at this aspect. As delivered, the LED is set to Mode 1. To move to the next mode (ie in this case Mode 2) you do the following:

1. Power-up the LED
2. Hold in the pushbutton
3. Wait until the LED flashes green
4. Release the button

Now here is the trickier bit. To confirm what mode you have now set, turn off power and then re-connect it. In this case you would expect to see three green flashes (indicates the LED is still in its 12V battery monitoring setting), a pause, and then green flashes (indicates Map 2).

The maps settings (the *second lot of flashes*) are as follows:

Map 1 – red flashes
Map 2 – green flashes
Map 3 – yellow flashes
Map 4 – red and then green flashes
Map 5 – yellow and then red flashes
Map 6 – yellow and then red flashes

Once you've sorted this aspect out, changing the voltage to 24V or 6V monitoring is straightforward. The procedure is:

1. Power-up the LED
2. Hold in the pushbutton
3. Wait until the LED flashes green, then red
4. Release the button

The voltage mode will switch to the next, so 12V to 24V to 6V to 12V – and so on. Then examine the *initial* flashes after switch-on on the basis of the following:

- 6V – red flashes
- 12V – green flashes
- 24V – yellow flashes

This little unit is ideal for battery monitoring in cars (especially those driven irregularly) and is very easy to fit.

VOLTAGE BOOSTER

Here's a compact voltage booster that will let you increase the output of the output of pumps, lights and fans! The module is cheap, has infinitely adjustable voltage output from 12-35V, and has a good power rating. In fact, it's rated at 150 watts (with added fan cooling), 100 watts without fan cooling (but maybe with bigger heatsinks), and I'd say on the basis of my testing, it will be very happy at 50-60 watts continuous ... just as it comes out of the packet. Search online under 'DC-DC 10-32V To 12-35V 150W Power Supply Boost.'

So if it will run items drawing up to about 50 watts, what can you use it for? Another way of expressing that wattage is to say that the standard continuous current draw of the device shouldn't be more than about 4A at 12V. In turn, that excludes high current fuel pumps that can easily draw up to 10A, and radiator fans that can take up 20A or more. So what can you run?

You can use the Voltage Booster to easily increase the pressure and output of windscreen washer pumps commonly used for intercooler water sprays. You can use it to boost the output of interior lights, brake lights and reversing lights. Depending on the measured current draw, you can use it to boost the flow of water/air intercooler

11. ELECTRONIC BUILDING BLOCKS

This little module will boost 12V battery voltage to as much as 35V, and can supply 50W loads (and more with added fan cooling).

pumps. You could also use it with 50W headlights in order to brighten your main beam (one unit per light).

Finally, you can use it to maintain the output of a nominally 12V lighting system, even as battery voltage falls. In testing, we found that the system would provide its boosted and regulated output down to an input voltage of 10V. To put that another way, there would be no change in light brightness from a fully-charged battery of 13.8V to a flat battery at 10V!

But what happens to the device you're powering when you increase the voltage going to it? In short, not only will its performance improve, but its life will be reduced. In many cases, that's of little concern – something like a water spray pump is used so little (in relative terms) that its life will still be fine. Incandescent light bulbs will have a shorter life, but as they're a replacement item, again it's no big deal. However, you should select the increase in voltage with care. A motor (eg a pump) used infrequently in short bursts could be run at 18V without many issues, but a filament lamp being used for long periods shouldn't be fed much over 15V. In general, make sure that items don't get too hot!

Note that devices that use internal voltage regulators, or are current limited, shouldn't be run at higher than normal voltages. Normally, there will simply be no difference in the performance of the device but in some cases (eg LEDs using dropping resistors), the device may be damaged by over-current. So LEDs and electronic bits and pieces like car radios and other electronic modules aren't suitable for running at higher than their design voltage.

The module comes as a built – but bare – circuit board. It's about 65 x 50 x 30mm (L x W x H), has a heatsink along each long side, a four-terminal connection strip at one end and a multi-turn pot at the other end. It's well made – in fact, a real quality item with clear connection markings (in English), good PCB design and four tapped metal spacers (on which it sits).

Connections are very easy – 'IN' positive and negative, and 'OUT' positive and negative. (Don't get these connections around the wrong way, and make sure you don't short-circuit the output.) Before powering-up, turn the pot many turns anti-clockwise to reduce the gain; turn it clockwise to slowly bring up the output voltage. The output voltage is easiest measured with a multimeter. When the wiring is complete, the board should be mounted in a ventilated box.

So how efficient is the module? Efficiency is important for two reasons. Firstly, the less efficient it is, the more heat it will have to dissipate. Secondly, the less efficient it is, the more energy it wastes. The manufacturer claims an efficiency of 94 per cent when running with an input voltage of 19V, a 2.5A current draw, and an output of 16V. That's not normally how you'd use it, though. In my testing, with an input voltage of 12.0V, an output of 14.7V and a current draw of 1.25A, the efficiency was 91 per cent. In other words, the power draw was 15.0W and the output power was 13.7W – an internal loss of 1.3W. That's pretty good.

HIGH CURRENT FLASHER

Here is a module that is ideal for running an emergency break-down flasher – or even just adding some pizazz at a car show. With the ability to drive loads drawing up to 7A, and with no less than 16 different flashing modes available at the touch of a switch, it's an impressive bit of gear.

While marketed as a device to flash your brake lights, the '12V-24V LED Brake Stop Light Lamp Flasher Module Flash Strobe Controller 16 Mode' module (search for it online) is useful for far more than just car brake lights.

The flasher comes packaged in a 58 x 35 x 16mm box – very compact indeed. At each end of the box are two wires – red and white. These are the input and output leads (red for positive). Marked on the box is an 'O' and arrow – this stands for 'output.' (If your box is not marked, the output end is that furthest from the DIP switch and closest to the output transistor.) Within the box, and accessed by lifting the lid, is a four-position DIP switch. This switch is used to select which of the 16 different flashing modes you want.

This tiny flasher module has 16 different modes and can handle loads of up to 7A. It's ideal as an alarm output operating lights or a siren.

WORKSHOP PRO   CAR ELECTRICAL AND ELECTRONIC SYSTEMS

Let's start with the most common use first: you want to flash a high-current LED as a warning. Connect the LED (complete with appropriate dropping resistor, and observing the correct polarity) across the output leads. Set all the DIP switches to 'O' (all the switches set closest to the output end) and then connect power (7-30V DC). The LED will flash at 18Hz (I measured 17Hz, but that's close enough). Set switch 1 to position 1, and the flash rate changes to 12.5Hz (I measured 11.8Hz). Note that you must disconnect power and then re-apply it for the change in switch position to take effect. Flash rates from the aforesaid 18Hz down to 1Hz are available, all with a 50 per cent duty cycle. These settings comprise modes 1-8.

The next range of modes (mode 9-12) are lower current consumption modes. Mode 9 turns on the output for 50ms once per second (a 5 per cent duty cycle), while Mode 10 flashes three 50ms pulses then stops for a second, before again starting the cycle. Mode 11 outputs three 50ms pulses, but this time then stops for 2 seconds. Mode 12 flashes ten 50ms pulses then stops for 2 seconds.

Modes 13-16 cause gradual changes in the LED output. The rate of brightening and darkening varies with the different modes – for example, Mode 13 varies brightness one up/down cycle every 3 seconds, while Mode 16 does four cycles over 11 seconds. Incidentally, this variation in brightness is achieved via Pulse Width Modulation (PWM) occurring at a frequency of 235Hz.

The beauty of this 'multi-mode' approach is that the flasher function can be very much tailored for the situation. For example, if current draw isn't an issue and you want to attract attention, then flash a high-power LED at 6.5Hz. But if you want minimal current draw, select a single 50ms flash every second. The 'output off' current draw of the module is only 7mA, so overall power consumption will be quite low.

But what if you want to flash incandescent lamps? The rapid available flashing rates (like 9.5Hz) and short pulses (50ms) won't work with incandescent lamps – the thermal inertia of the filaments means that they just won't respond fast enough. In that situation, the 1Hz, 50% duty cycle mode (Mode 8) can be used, and the 'varying brightness' modes (Modes 13-16) are very effective.

And what about current handling? The module is rated at 6A on incandescent loads and 7A on LED loads. And how well does the current handling stack up? Very well, in fact. It's most likely that you'll be running high current loads when using incandescent lamps, with the most 'current hungry' mode being the cycling modes of 13-16. I ran the module at 8A in Mode 16 with incandescent loading, and all was fine. On LED loads, 7A is an awful lot of lighting – I ran an automotive LED light bar (7.5A) and again the module coped fine.

This table shows all 16 different flasher modes that can be selected by setting the onboard DIP switch.

Mode	S1	S2	S3	S4	Output
1	0	0	0	0	18Hz flash
2	1	0	0	0	12.5Hz flash
3	0	1	0	0	9.5Hz flash
4	1	1	0	0	6.5Hz flash
5	0	0	1	0	4.5Hz flash
6	1	0	1	0	3Hz flash
7	0	1	1	0	2Hz flash
8	1	1	1	0	1 Hz flash
9	0	0	0	1	Cycling single 50ms flash, stopped 1s
10	1	0	0	1	Cycling three 50ms flashes, stopped 1s
11	0	1	0	1	Cycling three 50ms flashes, stopped 2s
12	1	1	0	1	Cycling ten 50ms flashes, stopped 2s
13	0	0	1	1	One up/down brightness cycle every 3s
14	1	0	1	1	Two up/down brightness cycles every 5s
15	0	1	1	1	Three up/down brightness cycle every 7s
16	1	1	1	1	Four up/down brightness cycle every 11s

VARIABLE VOLTAGE POWER MODULE

This module allows you to control the brightness of lights and the speed of fans. Hook it up to a DC fan, and you can control the fan speed via the supplied knob. Power filament light bulbs (like the dash lights in most cars) and you can control their brightness, steplessly and without flicker. Want to have a cheap variable power supply? Just add one of these to a car battery or old laptop PC plugpack.

The module is rated by the maker at 3A, quite a high current considering the board's diminutive dimensions. But in testing I found it was quite happy to run at this maximum current, with the supplied heat-sink getting only a bit warm. In fact, I short-termed increased current to 4.5A and nothing turned to smoke ...

The measured pulse width modulation frequency is 25kHz. That is, the output signal is turned on and off 25,000 times per second. This is high enough in frequency that if you are controlling small motors (like those used in fans), the motor windings cannot be heard 'singing.' Input voltage range is 6-28V – so it'll work on everything from a 6V old motorcycle through to a 24V truck!

11. ELECTRONIC BUILDING BLOCKS

This module allows you to control the speed or fans and the brightness of lights drawing up to 3A.

The board is about 51 x 33 x 16mm (L x W x H), with the knob protruding about 19mm. Connections are via a four-way terminal strip, with the correct connections written on the bottom of the board in English. Four mounting holes are provided. This is an excellent module that is cheap enough to use where previously you might have used a resistor or voltage regulator.

ADJUSTABLE VOLTAGE REGULATOR

Often in this book I have shown circuits powered from 5V rather than the nominal 12V of car systems. The reason I have done this is that most cars use ECUs that produce a 5V regulated output voltage to feed sensors such as the TPS, airflow meter, MAP sensor and so on. Often you can tap into this 5V supply if you need a low current 5V source for another application. But what if you want to provide 5V as a standalone source?

Cheapest and easiest is to buy an LM317 adjustable power module and then set it to 5.0V output. The LM317 is rated at up to 1.5A output, but will probably require a heatsink to achieve this. While a heatsink is not supplied on the module I bought, it is easily attached as the regulator tab is quite accessible. Note that even the shortest of durations of short circuits on its output will kill the module – so be careful!

To find the module, search online under 'Low Ripple Buck Linear Regulated Power Supply LM317 Module.'

An adjustable output power supply, and is ideal for providing a regulated 5V when that's required.

FOUR-CHANNEL CAR SOUND AMPLIFIER

Here is a four-channel car sound amplifier with a maximum output of 68W per channel. That's more than enough, even with relatively inefficient car speakers, to give you plenty of volume and punchy bass. Even better, if you're happy to do some of your own metalwork and you already have some hardware like fasteners and spacers, the cost is very low. Quality? Far better than the vast majority of similar power car amplifiers!

The completed four-channel, 12V build-it-yourself amplifier. The power supply module is on the left, and the two three-channel amplifier modules on the right. When the lid is placed on top, it locates the fan above the amplifier modules.

Starting points

The heart of the amplifier comprises four LM3886 ICs. This audio amplifier IC has been around for a while – it's an oldie but a goody. Each is capable of 68W into 4Ω at a maximum distortion of 0.1 per cent. However, rather than start with the bare ICs, I use two prebuilt, two-channel modules available on eBay. Note that the selected modules require a ±28V DC supply, rather than the AC transformer supply that most of these modules are configured for. Therefore, when sourcing these modules, ensure they look exactly as pictured. To find them, search under 'Assembled LM3886TF Dual channel Stereo Audio Amplifier Board 68W+68W 4Ω 50W*2 8Ω.'

Next up, you'll need a power supply capable of driving these modules. Previously, developing such a supply would have been expensive and time-consuming – but now one is available off the shelf. It's called '1PC Switching boost

WORKSHOP PRO CAR ELECTRICAL AND ELECTRONIC SYSTEMS

This view shows the six switching transistors used by the power supply module. The eBay module is supplied complete with the insulating washers and pads for heatsink mounting. However, you need to supply your own board mounting stand-offs.

One of the requirements was that the 270-watt amplifier be reasonably compact and light. The final item has a mass of 1.75kg (just under 4lbs) and dimensions of 250 x 140 x 75mm (10 x 5½ x 3in).

Power Supply board 350W DC12V to Dual ±20-32V for auto.' While the output of the power supply is ±32V as it arrives (the onboard pot allows adjustment), the LM3886 is happy with up to ±42V, so that's fine.

I also chose to use a fan for heatsink cooling and triggered it via another eBay module. This module is called '20-90°C DC 12V Thermostat Digital Temperature Control Switch Temp Controller New.'

Wiring

The electronics aspect of building the amp is very easy. The power supply board input GND, K and 12V terminals are connected as indicated – GND to chassis ground, and the 12V terminal directly to the positive of the car battery. Use a high-current fuse in this battery supply – eg 20 amps. The K terminal requires 12V to switch on the power supply – normally it's connected to the 'power aerial'

This tiny module triggers the amplifier cooling fan on the basis of the temperature selected with the DIP switches. The remote temp probe is located between two of the LM3886 amplifier ICs on one of the heatsinks.

output of the head unit. (Or if you don't have this, you could connect it to any ignition switched 12V supply.) The power supply board's output VCC+, GND and VCC- terminals are, respectively, connected to the (+), GND and (-) terminals of the two amplifier modules. The line level inputs from the head unit are connected to the IN amplifier module terminal blocks (observe correct polarity), and the speakers connected to the OUT terminal blocks (again observe correct polarity). And that's it for wiring!

Building

You need to provide plenty of heatsinking capacity – either by the use of substantial heatsinks, or by using smaller heatsinks but adding a fan. In actual use, the majority of heat is generated by the four LM3886 modules – despite appearances, the power supply module heatsinking requirements are more modest.

I wanted a compact box, so made one from aluminium sheet specifically to suit the required dimensions. The overall dimensions of the box were about 250 x 140 x 75mm. The heatsinks were formed by using 8mm thick aluminium plate for two walls of the box. A salvaged 12V fan was placed in the top panel (and there is a matching size hole in the bottom panel) and the fan is triggered by the temperature sensing module. The fan is set to turn on at 40°C.

The power supply module comes with the required insulating washers and collars for mounting the transistors to the heatsink, while the amplifier modules uses plastic encapsulated ICs and so do not require any extra insulation. To mount the boards, you'll need to provide insulated standoffs and screws, washers and nuts.

Rather than place connectors for the inputs and speakers on the box, I chose to directly wire these

11. ELECTRONIC BUILDING BLOCKS

This prebuilt stereo amplifier module is available on eBay, with two of the modules needed for this amplifier. The module requires at least ±28V DC to run and, in use, plenty of heatsinking is also required.

connections to the boards. These leads were run through rubber grommets that slide up appropriate channels when the lid is screwed into place.

Obviously, the type of housing you place the components in is up to you – you could even use a discarded car sound amplifier enclosure that incorporates its own heatsink. But remember, whatever approach you take, you'll need either quite substantial heatsinking or need to add a fan.

Results

The only financial outlay I had was for the eBay amplifier and power supply modules. And for that money, this is an unbeatable amplifier. The sound is excellent – better than commercial car sound amplifiers costing two or three times much, and so much better than the typical four channel amplifier built into a head unit that it's not funny!

ELABTRONICS MODULES

Those of you who have more familiarity with electronics might wonder why I haven't covered any of the many microcontroller modules that are available – Arduino, Raspberry Pi and so on. There are two reasons for that: firstly, if you're experienced in electronics, you're probably already using such modules, and secondly, for less experienced people, using one of these modules in a car can be a bit frustrating. Why? Well, such modules are not normally configured to run on 12V, and they often don't have outputs that can directly drive real loads. Furthermore, even simple programming requires at least some familiarity with coding.

However, I'd like to briefly cover two modules that address most of these concerns. Both are available from eLabtronics, an Australian company (see www.elabtronics.com). Over the last couple of decades I have often worked with this company to develop automotive uses for their electronics. (Just to make it clear, I don't benefit financially at all from the sales of the modules.)

eLabtronics has two different modules that are relevant to us.

1. Multi-Purpose Module

The Multi-Purpose Module (MPM) comprises a smart, user-configurable module. The MPM works on any voltage from 10-40V DC. It features:
- two multi-turn user-adjustable potentiometers (pots)
- a four-terminal screw-down connector
- a four-position DIP switch that is used to configure the MPM to provide different features
- a fuse
- a microcontroller
- a high-current output transistor, called a MOSFET

The four-wire connector is for power, ground, signal input and output. To put this another way, the maximum number of connections that need to be made to the board comprises just four wires!

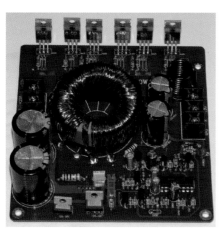

This power supply module generates ±32V DC from 12V DC and can be directly connected to the amplifier modules.

The eLabtronics Multi-Purpose Module (MPM) can directly drive high current loads via its MOSFET output transistor. It can be bought as a voltage switch, pulser or universal timer.

The hardware of the Multi-Purpose Module remains the same for all versions. The module is available in these different software versions:
- Voltage switch
- Pulser
- Universal timer

When the output MOSFET is activated, positive power is available on the output terminal. Without a heatsink fitted to the MOSFET, the module can output 3A. With a small heatsink that rises to 6A, and a larger heatsink will allow 10A to be drawn (of course, if you want to operate loads of greater than 10A, you can use an SSR on the output).

The two onboard pots are used to allow user-adjustment of two different parameters. (Some examples are covered below.) The DIP switch (not active on all versions of the MPM) allows the user to configure various options. The fuse protects against short-circuiting the output, and a diode is also fitted to protect against reverse polarity power connection. There's also a low current 5V supply pin on the board, allowing the powering of external temperature and light sensors.

The microcontroller is pre-programmed by eLabtronics to allow the MPM to perform the desired function. So what sorts of functions are available? Let's take a quick look, including some example uses.

eLabtronics Voltage Switch
- Switch on fans and pumps at certain temperatures
- Switch on lights as darkness falls
- Switch off battery chargers when battery is fully charged
- Low battery voltage warning

This single module will switch a power feed on (or off) when a monitored voltage reaches the adjustable trip-point. This allows the monitoring of temperature and pressure sensors, switching on the output when the levels rise too high or too low. In cars, engine management sensors like airflow meters and oxygen sensors can be monitored. You can also use this module to monitor battery voltage, turning off a charger when the battery is fully charged. Low battery warning alarms can also be easily implemented. One onboard pot allows the trip-point to be accurately set, and the other pot sets the difference between switch-on and switch-off points.

eLabtronics Pulser
- Flash high-powered LEDs
- Pulse horns and sirens

The Pulser is, as the name suggests, designed to pulse the output. The speed of the pulsing and the length of time the output is switched on each pulse are both independently adjustable by the pots.

This independent adjustment is key to the module's usefulness. You can have the output switch on for a very short time at a fast rate (think of rapidly flashing a high-powered LED), or you can have the output switch on for a longer time much less frequently (maybe pulsing a horn) – or the output can be on for a short time but at long intervals. You can even have the output switch on for 15 seconds every 3 minutes. As the examples suggest, in most cases the output MOSFET is powerful enough to directly drive lights, a horn or a pump.

The switch that's used to turn on the Pulser needs to carry only a tiny current, making it easy to trigger the Pulser from an existing alarm output or a low current switch like a temperature switch, light sensor or a pressure switch. The frequency can be set anywhere from 10 per second to once per hour, and the duty cycle can be set from 1 per cent to 99 per cent.

eLabtronics Timer
- Multi-purpose timer
- Many different modes
- Directly drive loads

The time length is set by the two pots and the timer also has many different modes, including one-shot, normally-on and normally-off.

In addition, most of the above modules also have additional functions – for example, the Voltage Switch output can be configured to give a couple of warning beeps, pulse the output on and off, or stay fully on. The Pulser can be triggered once a certain monitored temperature is reached. The beauty of the MPM is that it is well protected, can directly drive high-current loads, and is easily configurable (via the DIP switch and pots) to suit the requirements of the user. For full details, see the eLabtronics website.

2. Stemsel

The Stemsel controller module uses a PIC18F14K50 microcontroller. The module has:
- 12 digital or analog input/output pins
- four driver outputs that can directly drive relays (no need for protection diodes)
- an onboard USB connector that allows reprogramming from a PC or laptop
- 5V output that can be used to power sensors.

The module can be powered from the car's 12V supply.

The Stemsel module does not come pre-programmed to perform specific functions (with one exception – more on this in a moment), however an easy-to-use visual software system called CoreChart can be used to program it (see https://www.elabtronics.com/downloads.php). This programming approach is sufficiently flexible that lots of functions can be developed.

With its ability to be driven from a 12V source and its easy programming capability, this is a module that, while more expensive than generic modules, is a significant step easier for someone to program and use than other approaches.

And the exception – the pre-programmed function the

11. ELECTRONIC BUILDING BLOCKS

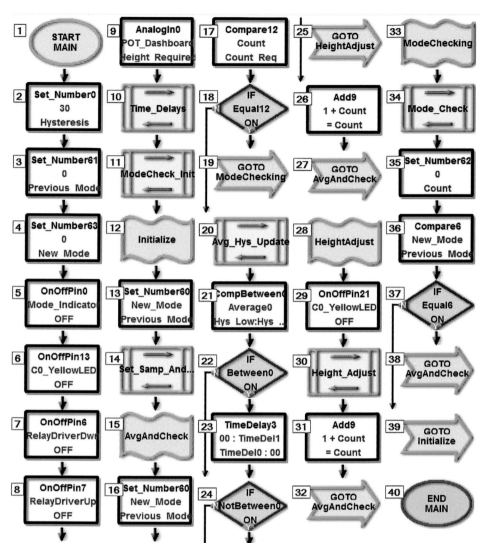

Here is an example of Corechart programming used on the Stemsel controller, in this case working as the eLabtronics Air Suspension Controller (eLASC). This is a complex program written by an engineering professional, but much simpler programs can also achieve excellent outcomes.

The eLabtronics Stemsel is a microcontroller module that will work off 12V and is programmable via the onboard USB connector. It can directly drive relays. Programming uses a simple visual language called Corechart.

module can be bought for? Stemsel is available as the eLabtronics Air Suspension Controller (eLASC), running purpose-developed software to control the height of two air springs. The eLASC is covered in detail in another of my books available from Veloce (see *Custom Air Suspension – how to install air suspension in your car on a budget!*) but in brief, the eLASC is set up to provide:

- Two relay outputs, allowing up/down control of a pair of air springs
- 5V regulated output to feed sensors and pots
- One suspension height sensor input, 0-5V
- One dashboard pot input for setting ride height, 0-5V
- One output for a monitoring LED
- A USB socket for making program changes (used only during tuning of set-up)

The eLASC also has quite sophisticated control logic – but that's another story!

159

Appendix

SAMPLE WIRING SYMBOLS

Shown on these two pages are wiring symbols used by Volkswagen on its wiring diagrams. Different manufacturers use different symbols, but these give you an idea of the types of symbols you should learn for the car on which you are working. (Courtesy Volkswagen)

Battery

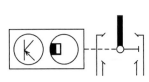

Distributor (Electrical)

Thermally operated switch

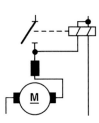

Starter

Spark plug connector and plug

Ignition coil

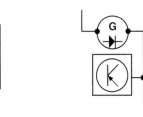

Generator

Glow plug heater element

Heater element (temperature dependent)

Push-button switch (manually operated)

Manually operated switch

Pressure operated switch

Mechanically operated switch

Multiple switch (manually operated)

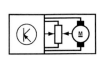

Control motor, headlight range adjustment

Digital clock

Antenna with electronic antenna amplifier

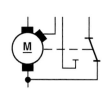

Wiper motor (two-speed)

Multifunction indicator

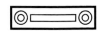

Radio

APPENDIX: SAMPLE WIRING SYMBOLS

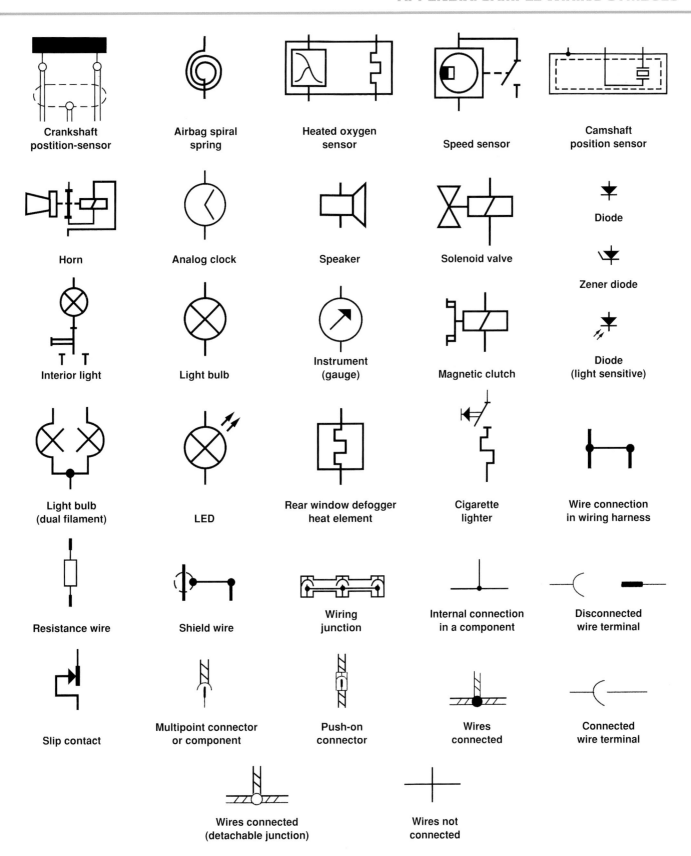

ALSO FROM JULIAN EDGAR

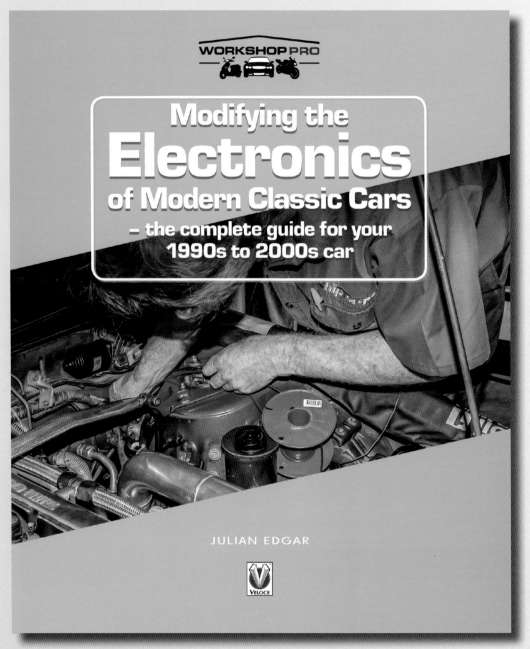

**Modifying the Electronics of Modern Classic Cars
– the complete guide for your 1990s to 2000s car**

Want to upgrade to fully programmable engine management but don't know how? Then this book has come to your rescue! Step-by-step instructions and 550 photographs and diagrams will empower you to upgrade your car's lighting, instrumentation, sound and engine management systems, and much more.

ISBN: 978-1-787113-93-0
Paperback • 25x20.7cm • 288 pages • 577 pictures

For more information and price details, visit our website at **www.veloce.co.uk**

MORE FROM VELOCE

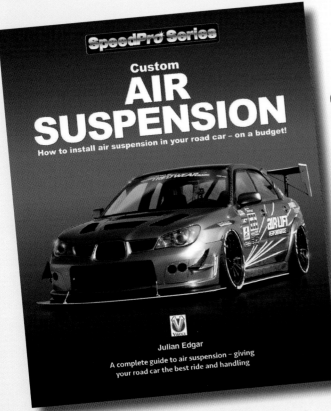

Custom Air Suspension
How to install air suspension in your road car – on a budget!

The first book that shows you how to fit air suspension to your car. It covers both theory and practice, and includes the step-by-step fitting of aftermarket air suspension systems and building your own with parts from other cars. If you want the best ride and handling for your road car, this is the book you need!

ISBN: 978-1-787111-79-0
Paperback • 25x20.7cm • 64 pages • 82 colour and b&w pictures

Setting up a Home Car Workshop

Want to modify, restore or maintain your car at home? This book is a must-read that covers the complete setting-up and use of a home workshop. From small and humble to large and lavish – this book shows you the equipment to buy and build, the best interior workshop layouts, and how to achieve great results.

ISBN: 978-1-787112-08-7
Paperback • 25x20.7cm • 160 pages • 250 pictures

email: info@veloce.co.uk • Tel: +44(0)1305 260068

ALSO FROM JULIAN EDGAR

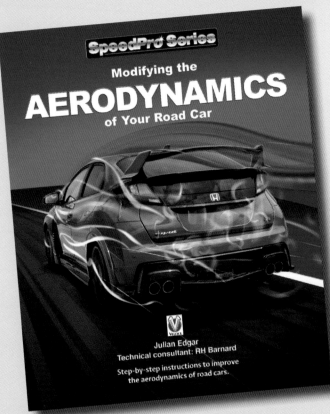

Modifying the Aerodynamics of Your Road Car

How to improve handling, straight-line performance or fuel economy. This handbook takes you from testing to making sophisticated aerodynamic modifications that have real impact.
ISBN: 978-1-787112-83-4
Paperback • 25x20.7cm

Optimising Car Performance Modifications

This book shows you how to easily measure on the road the gains and losses of changing air intakes, exhausts, cams and turbos. Also learn how to test suspension, brakes and car aerodynamics – accurately and at low cost.
ISBN: 978-1-787113-18-3
Paperback • 25x20.7cm

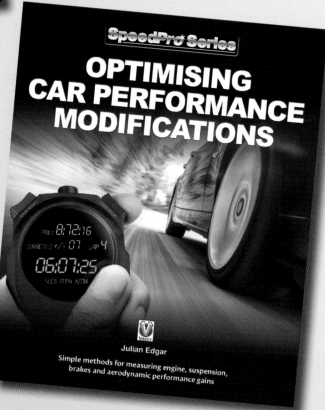

For more information and price details, visit our website at **www.veloce.co.uk**

MORE FROM VELOCE

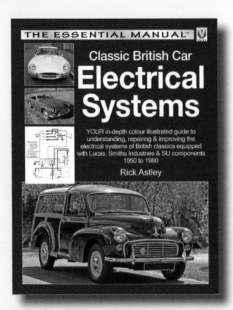

This book provides the theory, component parts, and full system operating explanations for each major electrical system used from 1950 to 1980, with particular emphasis on components that were ubiquitous in British cars of the period.
ISBN: 978-1-845849-48-1
Paperback • 27x20.7cm • 192 pages • 419 pictures

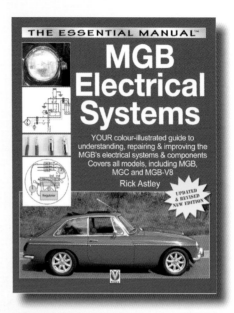

This book is essential reading for every MGB enthusiast. Each system in the car has its own chapter, with simple and uncluttered circuit diagrams in which each wire can be seen in its real colours.
ISBN: 978-1-787110-52-6
Paperback • 27x20.7cm • 192 pages • 400 pictures

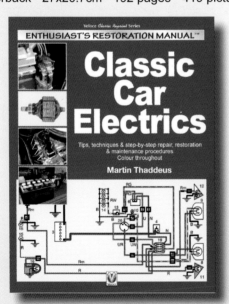

Reprinted after a long absence, this full colour book provides clear and complete information for the classic enthusiast who wishes to service, repair or improve car electrical systems.
ISBN: 978-1-787111-01-1
Paperback • 27x20.7cm • 96 pages • 301 colour pictures

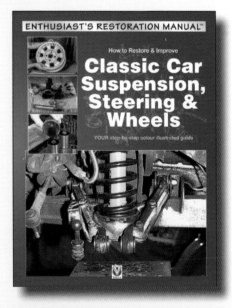

A step-by-step illustrated guide to restoring and improving the suspension, steering and wheels of your classic car, this easy to follow manual is fully illustrated, and packed with useful tips and techniques.
ISBN: 978-1-787111-87-5
Paperback • 27x20.7cm • 144 pages • 680 pictures

email: **info@veloce.co.uk** • Tel: **+44(0)1305 260068**

Index

ABS 126-128
Air suspension 124, 125
Airflow meter 52, 77, 86, 87
Alternator 42, 43
Amplifier 155-157
Amps 13-15
Analog signals 52
    measuring 58
Auto-stop 121
Automatic transmissions 131, 133

Battery 42, 44
Battery monitoring 151, 152
Bosch L-Jetronic 92-94
Bosch ME-Motronic 101-105
Bosch Mono-Jetronic 94-98
Bosch Motronic 98-101

Camshaft position sensor 78, 79, 83, 88
CAN bus 55-58, 82
Capacitors 67, 68
Charging systems 42, 43
Circuits 10-13
Climate control 130, 131
Closed loop 90, 91
Common rail diesel injection 111-117
Continuity 28
    measuring 33
Coolant temperature sensor 52, 78, 86, 120, 121
Cooling system control 120, 121
Crankshaft position sensor 88
Current 13
    measuring 33, 34
Current clamp 30
Current draw – starter 48, 81

Data bus signals 55-58
Digital signals 53
    measuring 58, 59, 82
Diodes 69
Direct petrol injection 105
Duty cycle 28, 34, 54, 90
    measuring 34

eLabtronics 157-159
Electronic Control Unit 90-92
Electronic stability control 128
Electronic throttle control 101
Engine management 86

Fault-finding 38, 136, 141
Flasher 153, 158
Freeze frames 140
Frequency 53-55, 58, 59
    measuring 34, 76
Fuel trims 91
Fuses 11, 12, 38-40

Ignition coil 48, 90
Ignition system 48, 49, 81, 82
Inductive sensor 78, 79
Injectors 80, 89, 90

K-line 82, 83
Kits – electronic 71
Knock sensor 79, 89

Lambda probe (see oxygen sensor)
Lean cruise 91
Limp home 91, 92
LIN bus 58, 59, 82, 83

MAP sensor 87, 88
MOST bus 58, 59
Multimeters 28-35
    buying 31
    leads 29
    logging multimeters 35

OBD 136-141
    codes 139
    readers 137
Ohms 13-15
Ohms Law 14, 15
Open loop 90
Oscilloscopes 74, 75
    specifications 75

INDEX

Over-run fuel cut-off 91
Oxygen sensor 88

Parallel circuits 10, 12
Period 76
Potentiometers 64
Pressure sensor 30
Pull-down resistor 63
Pull-up resistor 63
Pulse width modulation 54, 80

Relays 20-25
   boards 25
   solid state 24, 67
   specifications 23
Resistance 13-15
   measuring 32
Resistors 62
   parallel 63
   series 63
Rev limiting 91

Scan tools 138
Scope (see oscilloscope)
Seats – electric 142
Self-diagnosis 136
Series circuits 10-13
Short circuits 11
Sine wave 76
Square waves 76
Stability control 128
Starting systems 45, 47-49
Steering 129, 130

Switches 18-20
   circuits 19
   terminals 20
   types 20
Symbols – wiring 160, 161
Systems approach 124-126

Temperature controller 149
Throttle position sensor 89
Timers 148, 158
Torque control 103, 104
Transistors 70
Turbo boost control 119, 120

Units 15

Waveforms 76
   measurement 76
Wiring looms 46

Variable valve timing 118
Voltage 13
   battery 42, 43
   measuring 31, 77
   variable 154
Voltage booster 152
Voltage dividers 64
Voltage drops 45
   starter 47, 48
Voltage regulator 155
Voltage switch 150
Volts 13-15

NOTES